Methods of Motion

Revised Edition

An Introduction to Mechanics

Book One

Jack E. Gartrell, Jr.

◆

A Project of Horizon Research, Inc.

Materials for middle-grade teachers in physical science

This project was funded by BP America, Inc.

The National Science Teachers Association

Produced by Special Publications
National Science Teachers Association
1742 Connecticut Avenue NW
Washington, DC 20009

Stock Number PB 39
ISBN 0-87355-085-4

METHODS OF MOTION

Table of Contents

◆Acknowledgements

A project like *Methods of Motion: An Introduction to Mechanics, Book One* does not reach publication without the contributions and enthusiastic support of many people. Foremost among these are Iris R. Weiss, Project Director at Horizon Research, Inc., who initiated the program to produce this material, and Jack E. Gartrell, Jr., the principal author.

Karen Johnston provided the outline of concepts to be included, and Ellen Vasu developed the basic design for the modules. Elizabeth Woolard, John Steinbaugh, Larry Schafer, Norman Anderson, Ruth Sanders, and Charles Beehler provided valuable suggestions for activities and resources to use in developing this material.

The original manuscript was reviewed by Fred Goldberg, Robert Gioggia, Larry Schafer, and Judith Bodnick. They not only verified scientific accuracy, but evaluated the presentation of the concepts involved and the ease with which the activities could be performed. This revised edition of *Methods of Motion* was made possible through the efforts of Mario Iona, principle reviewer, and Larry Schafer and Jack E. Gartrell, Jr., who used their extensive classroom experience with the original manuscript to update, revise, and expand the activities and suggest changes which have improved the book as a whole.

Sondra Hardis of Standard Oil Company of Ohio provided many valuable suggestions for disseminating this material, helping to establish links between it and many potential users. Phyllis Marcuccio of the National Science Teachers Association handled the arrangements that made production of this book possible.

Methods of Motion was produced by NSTA Special Publications, Shirley Watt Ireton, managing editor; Michael Shackelford, assistant editor; Cheryle Shaffer, assistant editor. Michael Shackelford was NSTA editor for *Mechanics*. At AURAS Design, the book was designed by Sharon H. Wolfgang and production was handled by Ellen R. Baker. Illustrations were created by Sophie Burkheimer. Illustrations for Activity 18 are by Max-Karl Winkler. The contributions of all these people were essential in the preparation of this book.

Methods of Motion was a project of Horizon Research, Inc. Project Director was Iris R. Weiss; Project Co-director was Ann C. Howe.

Special thanks go to British Petroleum America, Inc., for providing the funds to make this project possible.

OVERVIEW

Methods of Motion
An Introduction to Mechanics

Mechanics is the branch of physical science that describes the behavior of systems under the action of forces. Much of our knowledge of mechanics is based on the work of Sir Isaac Newton. In 1687, Newton published a brilliant synthesis of the principles of mechanics—*Philosophiae Naturalis Principia Mathematica* (commonly known as *Principia*). Newton's insights were so powerful and complete that they remained virtually unchallenged and unmodified for two centuries. This set of modules is designed to illustrate and explain Newton's work in a way that will be useful to teachers who work with students in the middle grades.

Introducing mechanics in the classroom is often difficult. It sometimes seems that no matter what aspect of the topic the teacher presents, the students need advance knowledge of other information before dealing with the concept at hand. Beginning in the middle seems inescapable.

To further complicate matters, everyday observations of moving objects seem to contradict the unifying principles of mechanics. Newton's first law of motion states that an object in motion will continue in motion at a constant velocity in a straight line unless acted upon by an unequal force. But any sixth grader can tell you that when you throw a baseball, it soon curves to Earth and stops moving all by itself. No reasonable person believes that the ball really would keep moving in a straight line at a constant speed, no matter what the laws of motion may predict. All of our experience suggests otherwise.

◆Organization of this book

There is no perfect solution to the problem of how to begin studying motion. Neither is there a single best method to correct people's flawed assumptions about the behavior of objects that seem to disobey the laws of motion. The approach used in this series of modules is to illustrate selected concepts of mechanics with hands-on activities and audiovisual materials. Several segments of the *Eureka!* video series, produced by TVOntario, are used for concept presentations.

Following the modules is a collection of readings. This section gives detailed explanations of the concepts presented in the activities. These readings can be used to give the teacher some background of concepts to be explicated in the activities. They also can be reproduced as student handouts to supplement textbook materials or augment student discussion at the completion of the activities. Permission to reproduce both the readings and the activities for classroom use is given by the National Science Teachers Association.

A guide for teachers and workshop leaders is provided for use in planning instruction of the activities. This section lists equipment and materials required to perform each module's activities. Ordering information for the recommended audiovisual materials is also included in the guide for teachers and workshop leaders. Finally, a glossary provides definitions of many of the terms used in these modules. This glossary also serves as a master list of vocabulary introduced and defined in the activities.

◆Getting ready for classroom instruction

Most of the activities in this book are intended to be hands-on experiences in which the students can actively participate. Some activities function better as or *must* be performed as demonstrations, but even those intended as hands-on activities can be easily adapted for classroom demonstrations.

Each module begins with a discussion of the rationale, objectives, and overview of the activities within it. Each activity begins with a reproducible student worksheet, which also contains the concept objective and any new vocabulary words. Each activity is designed to take about one class period (40–60 minutes) to complete.

Each activity worksheet is followed by a commentary for teachers called "Guide to Activity...." These sections explain the expected results for the activity and provide additional background information on the content of the lesson. You also will find suggestions for time management, teacher preparation, ways to help students obtain reproducible results, ways of troubleshooting equipment, and hints about problems that may be encountered while performing the activity. Notes on safety, sample data, and answers to the questions on the activity worksheet are also provided. In addition, these commentaries contain suggestions for further study, outlining other experiments that can be performed using the same apparatus.

The equipment required for the activities consists mainly of inexpensive toys and other low-cost, readily available materials. Good results can be obtained for many activities by using alternative procedures or by using substitutes for materials listed on the worksheet. If you do not have ready access to the materials listed, the "Guide to Activity...." section offers possible substitutions and procedure modifications.

You will note that these modules use metric units wherever possible. These are the units routinely used in science. Length is typically measured in meters (m). One meter is equivalent to 39.37 inches, or slightly more than a yard. Shorter lengths are measured in centimeters (cm) or millimeters (mm). One meter is equal to 100 cm or 1000 mm. Large distances are measured in kilometers (km). One kilometer is equal to 1000 m, or about 0.6 mile.

Volume is typically measured in liters (L). One liter is slightly more than a quart. A gallon is equal to 3.8 L. Small volumes are usually measured in milliliters (ml). 1 L is equal to 1000 ml. One teaspoon of a liquid is equal to about 5 ml.

Time is typically measured in seconds (s). Other units familiar from the English system are occasionally used, but the second is the most common.

The metric units for mass, the kilogram (kg), and force, the newton (N), are discussed in depth in the modules and readings. 1000 grams (g) is equal to 1 kg.

In many cases the English system equivalent is provided to help you and your students make the translations. We have also included a metric conversion chart at the end of these modules.

◆Getting ready for workshops

If you are using this book as a workshop leader or teacher participant, you may wish to perform some of the activities as classroom demonstrations. As you work through these activities with your colleagues, you will have the opportunity to discuss new insights, explore alternative procedures, and make note of any problems you encounter. These experiences will add to your confidence when you direct your students in similar activities later.

Each module is designed to take one inservice workshop time period. The general topic of each module is described in its introduction. Each introduction lists instructional objectives for the workshop, gives the titles of the activities in the module, and indicates the readings that should be studied after the module's activities are performed.

MODULE 1

Sizing it Up

◆Introduction

•Why do astronauts seem weightless while orbiting the Earth?

•What determines how far an Olympic athlete can jump?

•How do race car drivers use lap times to determine their speed?

Answering these questions requires using the principles of mechanics. It also requires the measurement of three fundamental quantities: length, mass, and time.
The activities in Module 1 provide practice in designing experiments, making accurate measurements of moving objects, and recording and analyzing these measurements.

◆Instructional Objectives

After completing the activities and readings for Module 1, you should be able to

- measure distance using metric units [Activity 1 and Reading 1]
- determine the period of a pendulum [Activities 1 and 2]
- identify the control for an experiment [Activity 2 and Reading 2]
- identify variables in an experiment [Activity 2 and Reading 2]
- measure time intervals using a timer other than a standard watch or clock [Activities 3 and 4]

◆Preparation

Study the following readings for Module 1:

Reading 1: Measurement Skills Used in the Study of Motion

Reading 2: Identifying Experimental Variables and Controls

◆Activities

This workshop includes the following activities:

Activity 1: Measuring the Period of a Pendulum

Activity 2: What Can Change the Period of a Pendulum?

Activity 3: Building a Drip Timer

Activity 4: Now Wait Just a Minute!

ACTIVITY 1 WORKSHEET

Measuring the Period of a Pendulum

Materials

Each group will need

- a meter stick
- a piece of string 1.5 m long
- a paper clip
- a 25–50-g mass (A large fishing sinker may be used. 2 ounces is about 50 g.)
- a support for the pendulum (a ring stand or some place to tie the string so that the pendulum bob can swing freely)
- a watch or clock with a second hand

Vocabulary

- **Cycle (of a pendulum):** One of the "out and back" swings of a pendulum.
- **Gravity (force of gravity):** The force of attraction between all objects in the universe. Wherever there is mass, there is gravity.
- **Mechanics:** The branch of physical science that describes the behavior of bodies in motion; mechanics deals with energy and forces and their effects on bodies.
- **Pendulum:** Any object attached to a fixed point so that it swings freely back and forth under the action of gravity.
- **Period (of a pendulum):** The time it takes a pendulum to swing out and back to its point of release.

◆Background

A **pendulum** is any object attached to a fixed point so that it swings freely back and forth under the action of **gravity**. Studying the motion of a pendulum is one way to learn about the principles of **mechanics**. The **period** of a pendulum is the time it takes to swing out and back to its point of release. One of these "out and back" swings is a **cycle**.

◆Objective

To look at techniques of measurement and timing using a simple pendulum that you assemble

◆Procedure

1. Tie the paper clip onto the end of the string. Open one end of the clip slightly and hook the weight onto the paper clip (see top diagram).

2. Set up your pendulum as shown in the lower diagram. Adjust the

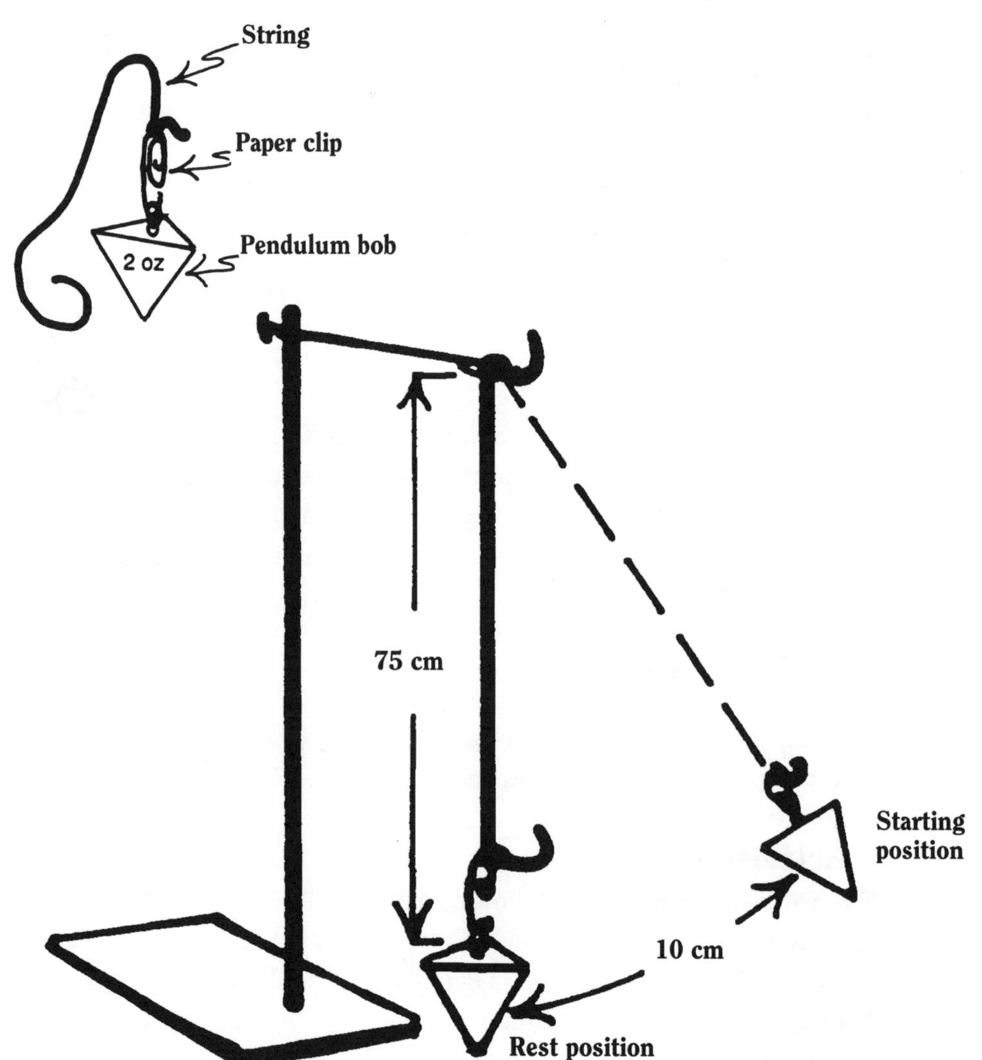

length of the string so that the bob hangs motionless 75 cm beneath the support. This is the "rest position."

Test the pendulum to be sure that it can swing back and forth freely. Pull the bob back about 10 cm from the rest position to the "starting position." Release it and time how long it takes to complete 10 cycles (10 complete swings back and forth).

Record your observations in the data table. Repeat your measurement two more times.

Data table

Trial	Number of cycles	Seconds required for 10 cycles	Period of pendulum= seconds for 10 cycles ÷ 10
1	10	_____ s	_____ s
2	10	_____ s	_____ s
3	10	_____ s	_____ s

3. The **period** of a pendulum is defined as the time that it takes to complete one cycle—that is, to swing out and to return to the point of release.

Calculate the period of the pendulum for each of your measurements by dividing the time required for 10 cycles by 10. Enter your results in the data table.

4. Compare the three values that you calculated for the period of the pendulum. The results should be almost equal for all three trials. If one of the periods you calculated differs from another by half a second or more, repeat your observations and recalculate the period of the pendulum.

5. Describe a way to use your pendulum to time an athletic event such as a 400-meter run.

____

____

____

____

6. Why do you think that judges of athletic events use stopwatches or other timers rather than a pendulum for timing races?

____

____

____

____

GUIDE TO ACTIVITY 1

Measuring the Period of a Pendulum

◆What is happening?

The purpose of this activity is to determine the period of a simple pendulum. The activity requires accurate time determination and the use of metric units for distance measurements. With a little care, a very reliable value for the period of the pendulum can be obtained. This is important, because the next activity will require accurate determinations of the periods of several different pendula.

You will find that all measurements in any experiment are subject to error. Error can result from the equipment used to measure, from the person doing the measuring, or from any of countless other sources. A measurement can never be completely free of error, but the experimenter tries to minimize error as much as possible.

◆Time management

One class period (40–60 minutes) should be enough time to complete the activity and discuss the results.

◆Preparation

Almost anything can serve as a pendulum bob, but metal washers or fishing sinkers (1 ounce or larger) are particularly good choices. A 1-ounce sinker has a mass of approximately 25 g. Air resistance can affect the period of a low-mass pendulum, and large or irregularly shaped pendulum bobs may not swing smoothly. Remind students not to pull the weight back more than 10 cm from the rest position. Longer swings can affect the period. The pendulum used in this activity should be saved for use in Activity 2, when it will serve as a basis for comparison to other pendula.

◆Suggestions for further study

Watch a pendulum slow down as it comes to rest. Does the period become shorter, longer, or stay the same? Find the answer by determining the period of the pendulum after it has slowed down but is still swinging back and forth. Pendula usually keep the same period as they slow down.
Can you think of other periods and cycles? How about the water cycle? What is the period for a water cycle? In other words, how long would it take for a speck of water to evaporate from the ground, go into the air, and return to the ground in the form of precipitation? What is the period for the human life cycle? A guitar string making a middle C tone cycles back and forth about 264 times in a second. If we want to know the period, or the time it takes the string to make one cycle, we just divide one second by 264. Since there are 264 cycles in one second, it takes 1/264 of a second for one cycle.

◆Answers

3. 10 cycles of a 75-cm pendulum require about 16 seconds. Therefore, the period of the pendulum is about 1.6 seconds.

5. Count the number of cycles of the pendulum during the race. If the period of the pendulum is known, the runners' times can be calculated by multiplying the period of the pendulum by the number of cycles. If the period is not known, the unit "one cycle of the pendulum" could be used as a standard measure of time in the same way in which we use the minute.

6. Stopwatches and timers are more convenient and much more accurate.

ACTIVITY 2 WORKSHEET

What Can Change the Period of a Pendulum?

Materials

Each group will need

- the pendulum and other materials used for Activity 1
- 2 additional pendulum bobs (use a 1-ounce [25 g] and a 2-ounce [50 g] fishing sinker)

Vocabulary

- **Control:** An experiment (or factor in an experiment) that remains unchanged for purposes of comparison. The control allows an experimenter to judge the effects of changing factors in an experiment.
- **Variable:** A factor that is changed in an experiment in order to see how it affects the results of the experiment.

◆Background

When performing experiments, it is often important to have one set of results that can be used as a standard against which other results can be compared. A **control** is an experiment (or factor in an experiment) that remains unchanged for purposes of comparison to all other observations in an experiment. In this activity, you will change the way that your pendulum is set up in order to investigate what determines the period of a pendulum. A factor that you change for an experiment is called a **variable.** The variables for this activity are the mass of the pendulum and the length of the string.

◆Objective

To determine how different variables affect the period of a simple pendulum

◆Procedure

1. Determine how long it takes a 50-g pendulum bob suspended from a string 75 cm long to complete 10 cycles. For each of three trials start the pendulum by pulling the bob back about 10 cm from the rest position. Calculate the period for each trial, and then calculate the average period for the three trials.

Record your results for this "standard bob" in the data table.

2. Remove the standard bob from the paper clip hook. Replace it with a 25-g mass. Start the pendulum by pulling the bob back 10 cm and releasing it, and time how long it takes this "light bob" to complete 10 cycles. Record your observations in the data table.

3. Remove the light bob from the paper clip hook. Place two 50-g masses on the paper clip. Time this "heavy bob" as you did the others, and record your observations in the data table.

4. Remove the heavy bob from the paper clip and replace it with the standard bob. Untie the string from the support and lower the bob so that it hangs 100 cm below the support. Time this pendulum, and record your observations in the data table under the heading "standard bob, long string."

5. Shorten the string so that the bob hangs 50 cm below the support. Test this pendulum, and record your observations under the heading "standard bob, short string."

Data table

Trial	Type of pendulum	Mass of bob	String length	Time for 10 cycles	Period (s)	Average period
1	standard	standard	75 cm	_____ s	_____	
2	(control)	standard	75 cm	_____ s	_____	
3		standard	75 cm	_____ s	_____	____
1	light bob,	light	75 cm	_____ s	_____	
2	standard string	light	75 cm	_____ s	_____	
3		light	75 cm	_____ s	_____	____
1	heavy bob,	heavy	75 cm	_____ s	_____	
2	standard string	heavy	75 cm	_____ s	_____	
3		heavy	75 cm	_____ s	_____	____
1	standard bob,	standard	100 cm	_____ s	_____	
2	long string	standard	100 cm	_____ s	_____	
3		standard	100 cm	_____ s	_____	____
1	standard bob,	standard	50 cm	_____ s	_____	
2	short string	standard	50 cm	_____ s	_____	
3		standard	50 cm	_____ s	_____	____

6. Which pendulum had the longest average period? ____________________

7. Which pendulum had the shortest average period? ____________________

8. Which variable seems to be most important in controlling the period of the pendulum? ____________________

9. Which variable seems to have very little effect on the period of the pendulum?

____

10. What is the purpose of changing only one variable at a time?

____

____

____

____

GUIDE TO ACTIVITY 2

What Can Change the Period of a Pendulum?

◆What is happening?

The purpose of this activity is to determine how several variables affect the period of a simple pendulum. The activity requires the use of several scientific processes: identifying variables, modifying the conditions of an experiment, and analyzing data. The activity also provides practice in making measurements and determining time intervals accurately.

The major concepts that are introduced are **control** and **variable**. For scientific experiments, the data obtained for the control serve as the basis of comparison for other observations. The standard pendulum, set in motion by pulling the bob back 10 cm each time, uses a 50-g mass and a 75-cm string. It serves as the control for the activity.

The variables tested in this experiment are the mass of the pendulum bob and the string length. Other factors are not changed. There are many possible variables that are *not* tested in this activity—for example, the point of release, the color of the pendulum bob, the thickness of the string, the time of day, and the gender of the experimenter. No experiment is likely to test all of the possible variables that may or may not affect the results.

By systematically modifying the pendulum, studying one variable at a time, and comparing the results with the period of the standard pendulum, most experimenters will reach the correct conclusion: the length of the string supporting the bob is the variable that determines the period of the pendulum. The longer the string, the longer the period of the pendulum. The shorter the string, the shorter the period of the pendulum.

◆Time management

One class period (40–60 minutes) should be enough time to complete the activity and discuss the results.

◆Preparation

Remember to save the pendulum and materials from Activity 1. As mentioned in Activity 1, almost anything can serve as a bob, but metal washers or fishing sinkers are particularly good choices. As before, remind students not to pull the bob back more than 10 cm from the rest position.

◆Suggestions for further study

Systematically vary the length of the string for the pendulum. Make a graph of the period plotted against the string length. **Note:** This is *not* a linear relationship! The graph will form a smooth curve.

Construct and test different types of pendula. For suggestions about setting up unusual pendula, see "The Amateur Scientist" in the October 1985 issue of *Scientific American*.

Construct a large pendulum and see if its period changes when the bob is released from different heights above the rest position.

◆Answers

6. The standard bob on the long string had the longest period.

7. The standard bob on the short string had the shortest period.

8. The length of the string controls the period.

9. The weight of the pendulum bob has little effect on the period.

10. If more than one variable is changed at a time, there is no way to determine which variable is the cause of any changes in the period of the pendulum. When all other variables are kept the same while only one variable is studied, any changes in results have to be linked to the one variable.

ACTIVITY 3 WORKSHEET

Building a Drip Timer

Materials

Each group will need

- an empty bottle of dishwashing liquid and its push-pull top
- a timer or clock with a second hand
- a metal pie pan or other shallow metal pan
- a container of water (enough to fill the detergent bottle)
- heavy-duty scissors
- a ring stand and clamp to hold the bottle (optional)

Vocabulary

- **Calibrate:** To adjust a measuring instrument so as to reduce error in measurement; to standardize the instrument by determining its deviation from an accepted value.
- **Hypothesis:** An educated guess that is based on available information and experience. It is tested by scientific experiment.

◆Background

One way scientists investigate the world around us is to make guesses about the way things work and then design experiments to test those guesses. Scientists don't guess wildly or randomly though; they base their ideas on information and experience and then make an educated guess about the outcome of an experiment. This kind of educated guess, which is tested by scientific experiment, is called a **hypothesis**. In this activity, you will use one means of measuring time and test a hypothesis based on data you collect in the experiment.

◆Objective

To build and **calibrate** your own drip timer, then to use it to test a hypothesis

◆Procedure

1. Place the pie pan on the floor under the edge of the table. Cut the bottom off of the dishwashing liquid bottle. If you have a ring stand, place the bottle in the stand so that its top hangs beyond the edge of the table directly above the pie pan. If you do not have a ring stand, have a member of your group hold the bottle at the edge of the table above the pie pan on the floor. Fill the bottle with water.

2. Open the bottle top until water comes out at a rapid, steady drip. Position the bottle so that the pie plate catches the drops of water. If the water comes out in an unbroken stream or drops slowly and irregularly, adjust the push-pull top down or up a little and try again. Continue adjusting the opening of the top until you can get fast, even dropping rates. **Note:** This process involves trial and error. Be systematic and try to move the closure only a tiny distance each time you adjust it. Once you get fast, regularly spaced drops, be careful not to move the sliding closure!

3. After the timer is dripping at a fast, regular rate, see how long it takes for it to drip 30 times. Fill in the following data table:

Data tables

Trial	Seconds for 30 drops to fall
1	_____ s
2	_____ s
3	_____ s
4	_____ s
5	_____ s
	_____ **total seconds** (trials 1–5)

4. Use your data for the 5 trials to calculate the average number of seconds required for 30 drops to fall into the pan.

   Average time per 30-drop trial = total seconds ÷ 5 = _____ s

5. Use your answer for question 4 to calculate the average number of drops per second.

   Average number of drops per second = 30 drops ÷ Average time per 30-drop trial = _____ drops per second

6. How many drops would fall from your timer during a 10-s interval? You could just guess how many, but now you do not have to guess. You can predict the result on the basis of your previous observations. Multiply your average drops per second (from question 5) by 10 s to get a *hypothetical* number of drops that will fall in 10 s.

_____ drops per second X 10 s = ______ drops in 10 s

On the basis of this calculation, complete the following statement of a **hypothesis**:

During a 10-s time interval, ______ drops will fall from the drip timer.

7. Now test your hypothesis by making the following observations:

Data table

Trial	Drops in 10 seconds
1	_____ drops
2	_____ drops
3	_____ drops
4	_____ drops
5	_____ drops
	_____ **total drops** (trials 1–5)

8. Calculate the average number of drops that fall during a 10-s interval.

Total drops ÷ 5 = _______ average number of drops in 10 s

9. Is the average number of drops that you *observed* in 10 s about the same as the number of drops that you *predicted* in your hypothesis? Discuss why your hypothetical value and your observations of drops falling during a 10-s interval are similar or different. Can you suggest ways to improve the accuracy of your drip timer?

____

____

____

____

____

GUIDE TO ACTIVITY 3

Building a Drip Timer

◆What is happening?

This activity illustrates another means of measuring time. It also requires participants to perform a common and important scientific task: calibrating a piece of equipment. Some degree of error in measurement is inevitable, no matter what type of equipment is being used. However, if a scientific apparatus is properly assembled and modified systematically, the variability of the experiment will be reduced.

Calibration is a major concern in all measurement. One way to calibrate an instrument is to compare its results to those of a similar, accurate instrument and make adjustments accordingly. Another way is to use the instrument to measure a known quantity and make adjustments based on the difference between the known value and the observed value.

In this activity, the instrument being used does not measure time in standard units and so does not need to match other devices, but it still needs to be calibrated. The water should drip at a steady, regular rate. By adjusting the closure and comparing a prediction based on the timer's past performance to new observations, the timer can be calibrated to ensure that it gives consistent readings.

"Building a Drip Timer" also gives a step-by-step illustration of how preliminary observations (in this case, setting up and systematically testing the dripper) lead to the formulation of a **hypothesis** for the second part of the activity. This illustration emphasizes the important fact that hypotheses are not haphazard or random. Students make observations, form a hypothesis, and then test this hypothesis with further observations.

◆Time management

One class period (40–60 minutes) should be enough time to complete the activity and discuss the results.

◆Preparation

Students can bring the detergent bottles and tops used in this activity from home. Save the bottles and tops for use in later activities.

Some brands of dishwashing liquid have tops that are much easier to adjust than others. It is a good idea to test them ahead of time.

One source of systematic error is built into this type of timer: as the water drains down, the rate of dripping tends to slow down. One way to prevent this is to add small amounts of water to the timer to replace what is dripping out.

◆Suggestions for further study

Many different cultures have used water clocks. Have your students build one of these historical models or invent one of their own.

◆Answers

The answers to these questions are *completely dependent on the rate at which the timer drips*. Rates may range from about 1 drop per second to around 5 drops per second. The sample data set that follows is included *for reference only. These answers are not the only correct and acceptable answers for this activity.*

3. The following table represents one possible set of observations:

Sample data

Trial	Seconds for 30 drops to fall
1	9.5 s
2	10.0 s
3	10.5 s
4	10.0 s
5	10.0 s
	50.0 **total seconds** (trials 1–5)

4. Average time per 30-drop trial = 50 s ÷ 5 = 10 s

5. Average number of drops per second = 30 drops ÷ 10 s = 3 drops per second

6. If the calculations of the number of drops per second are done correctly, the hypothesis about how many drops will fall in 10 s should (approximately) match the observed value.

The *hypothetical* number of drops that will fall in 10 s is:

3 drops per second X 10 s = 30 drops in 10 s

So, the complete statement of the hypothesis reads as follows:

During a 10-s time interval, 30 drops will fall from the drip timer.

7. The following table represents one possible set of observations:

Sample data

Trial	Drops in 10 s
1	28 drops
2	30 drops
3	32 drops
4	31 drops
5	29 drops
	150 **total drops** (trials 1–5)

8. Average number of drops during a 10-second interval = 150 ÷ 5 = 30 drops in 10 s

9. Differences between the observed and predicted values may be accounted for in several ways: error in counting a large number of drops is possible; the person timing the interval may not indicate exactly when 10 s have passed; the hole may be partially blocked for a while (or slip); other random errors are also possible.

ACTIVITY 4 WORKSHEET

Now Wait Just a Minute!

Materials

Each group will need

- an unconventional timekeeping device: this may be a drip timer, a pendulum, or another timer of your own design
- a time judge with an accurate clock or watch with a second hand and paper to record elapsed times

Vocabulary

- **Error in measurement:** The difference between an observed or calculated value in an experiment and the true value. An error in this sense is not a mistake, but a variation from the true value present to some degree in all experiments.

◆Background

An important part of the scientific process is minimizing the **error of measurement** for an experiment. In this activity, while measuring time with timekeeping devices you have already built or new ones you create, you will be attempting to minimize the error in your measurements.

◆Objective

Using a timekeeping device of your own design, determine an interval of exactly one minute. You may not refer to any clock, watch, or other commercially produced timekeeper while performing this task.

◆Procedure

1. You will be allowed 10 minutes to set up your unconventional timepiece. While setting up your timepiece you may use a watch or clock to calibrate it and minimize the error in its operation.

2. After the timepieces have been completed, the time judge will check to be sure that all watches and clocks are out of sight, and that all groups are ready to start their timers.

3. The time judge will count down 10 s, then give the command, "Start." The judge should immediately begin timing a one-minute interval, and be prepared to write down the actual elapsed times for each group participating in the activity.

4. On the "Start" command, each group uses its unconventional timer to determine a one-minute interval as accurately as possible. When the group believes that one minute has elapsed, a group member should say "Time" to the time judge.

5. The time judge should record the *actual* elapsed time for each group when that group calls out "Time." The group coming closest to an exact one-minute interval wins the competition.

GUIDE TO ACTIVITY 4

Now Wait Just a Minute!

◆What is happening

"Now Wait Just a Minute!" is designed to provide an enjoyable way to use the knowledge of timing gained in previous activities. There is no perfect solution to the challenge stated for this activity. By modifying and manipulating the pendula and/or the drip timer, or by devising a completely different type of time measuring device, groups performing this activity engage in an important scientific process: *minimizing the error of measurement for an experiment.*

Some error of measurement is inevitable, but scientists strive to reduce experimental error to the lowest level that is practical. Galileo used water clocks for some of his experiments; now some scientists routinely use timers that are accurate to one-millionth of a second.

◆Time management

Depending on the number of rounds of the competition, this activity could take only a few minutes or the entire class period. Rather than limiting students to a single class period to construct their timers and compete, you may wish to assign the activity as a "mini project" to be built outside of class. The completed timers can then be tested in class, and prizes given for the most accurate timer, most creative, best looking, etc.

◆Preparation

There are no "right" or "wrong" solutions to this activity. It is designed to get people thinking about how we measure time, and to consider how large the errors of measurement may be for any method of timekeeping. Having students design and build their timers outside of class encourages creative solutions and makes the best use of the class time available.

◆Suggestions for further study

There is an almost endless number of ways to vary this basic task. Encourage your students to be creative! Here are some possible changes in the rules for the activity:

- If you have a large number of groups with timers, you may wish to have only 2 or 3 groups at a time go "head to head" measuring a one-minute interval. This reduces confusion and makes the time judge's job easier.
- If determining a one-minute interval seems too easy, have the groups determine a longer interval, or have the judge announce "Start" and "Stop" and challenge each group to determine (in seconds) how long the unknown interval (determined by the judge) lasted.

MODULE 2

Mass and Force

◆Introduction

• Why is it harder to pick up a brick than a block of wood the same size?

• How does a parachute slow the fall of a skydiver? Does the skydiver's weight affect the rate of fall?

• Does a seemingly weightless satellite orbiting the Earth still have mass?

Students of Newton's laws of motion soon discover that everyday definitions of mass and force are not sufficient for analyzing complex motions. Correctly answering the questions listed above requires understanding the scientific definitions of mass and force.

The activities in Module 2 introduce the concept of inertial mass, provide practice in measuring force, and show how vectors are used to represent forces. The definitions of mass and force lead to a formal statement of Newton's first law of motion.

◆Instructional Objectives

After completing the activities and readings for Module 2, you should be able to

- explain the relationship between inertia and mass [Activities 5, 6, and 7]
- state three definitions of mass [Activity 6 and Reading 6]
- measure mass using an inertial balance [Activity 7]
- state Newton's first law of motion [Reading 3]
- define force [Reading 4]
- use vectors to represent forces [Activity 8 and Reading 5]

◆Preparation

Study the following readings for Module 2:

Reading 3: Concepts of Dynamics: Newton's First Law of Motion
Reading 4: Defining Force
Reading 5: Using Vectors to Represent Forces
Reading 6: Three Definitions of Mass

◆Activities

This module includes the following activities:

Activity 5: *Eureka!* #1—Inertia
Activity 6: *Eureka!* #2—Mass
Activity 7: Defining Mass as "Difficult-to-Moveness"
Activity 8: Applying Force to a Rubber Band

ACTIVITY 5: VIDEOTAPE

Eureka! #1—Inertia

◆Background

Eureka!, produced by TVOntario, is a series of animated videos using examples drawn from everyday experience to demonstrate the behavior of matter in motion. The presentations are an enjoyable way to review the concepts of mechanics. You may wish to use them in the classroom to help students develop an intuitive feel for the principles being studied.

The first segment of *Eureka!* partially defines **inertia**. It uses stationary objects on Earth and objects moving through air or floating in the **vacuum** of space to illustrate the tendency of matter to "keep doing what it is already doing."

◆Time management

The running time of the videotape is 5 minutes. At least 15 minutes should be allotted to introduce, run, and discuss the videotape. You may wish to play the videotape at the end of a lesson to reinforce the concepts presented.

◆Comments on the videotape

The videotape points out that if a pebble were kicked in the vacuum of space, it would continue traveling through the universe forever. You may wish to use this idea to explain to your class why the Voyager and Pioneer spacecraft will continue moving out into the universe after they are no longer being pushed ahead by their rockets.

Concept summary

"Things like to keep on doing what they are already doing. They don't like to start moving or stop moving. They are lazy. Another word for laziness is **inertia.**"*

Vocabulary

- **Inertia:** A property possessed by all matter that can be thought of as laziness or "difficult-to-moveness;" it is the tendency of matter to keep doing what it is already doing. Inertia is the subject of Newton's first law, which states: an object at rest tends to stay at rest, and an object in motion tends to stay in motion in a straight line and at a constant speed unless acted upon by an unequal force.
- **Vacuum:** A space devoid of all matter, including air.

*Eureka! Produced by TVOntario © 1981.

ACTIVITY 6: VIDEOTAPE

Eureka! #2—Mass

◆Background

Eureka! #2 refines the definition of inertia. The idea that inertia depends on the **mass**, or *quantity of matter* in an object , and does *not* depend on the *size* of the object is developed by comparing the ease of moving a large styrofoam cube with the difficulty of moving a much smaller lump of lead. Mass is equated with inertia. The concept that mass can be measured by using a two pan balance to compare the object to a standard cylinder of platinum alloy stored in Sèvres, France, is also illustrated.

Concept summary

"Inertia means laziness—the tendency things have to keep on doing what they are doing. Big things aren't always lazier than small things. It all depends on how much mass they have. The more mass, the more inertia."*

◆Time management

The running time of the videotape is 5 minutes. At least 15 minutes should be allotted to introduce, run, and discuss the videotape. You may wish to play the videotape at the end of a lesson to reinforce the concepts presented.

◆Comments on the videotape

The representation in this segment of the "pebble" on the right arm of an equal arm balance is somewhat deceiving. The pebble is shown being balanced by a 1-kg weight. What most people would call pebbles would probably weigh less than 1/20 kg (< 50 g). A 1-kg pebble would be called a rock, and would have a different volume in comparison to the standard kilogram mass which is illustrated.

A similar type of error exists in the representation of the platinum standard kilogram. Pure platinum has a density of 21.4 g per cubic centimeter. Lead's density is only 11.3 g per cubic centimeter, so the platinum kilogram would only be about half the volume of a lead kilogram. Common laboratory weights are often made of brass, having a density of about 8.5 g per cubic centimeter. A 1-kg brass weight would be about 2.5 times larger in volume than the platinum kilogram.

Despite the slightly misleading size relationships in the illustrations, the concepts are accurately presented in this segment.

Vocabulary

- **Mass:** A property of matter related to inertia. As the mass of an object increases, so does its inertia. Mass can be thought of as the quantity of matter in an object.

**Eureka!* Produced by TVOntario © 1981.

ACTIVITY 7 WORKSHEET

Defining Mass as "Difficult-to-Moveness"

◆Background

The amount of **mass** an object has is directly proportional to the object's **inertia**. The relationship between mass and inertia is an important one in mechanics; in fact, mass can be measured by making use of this relationship.

◆Objective

To build and use a mass-measuring device called an **inertial balance**

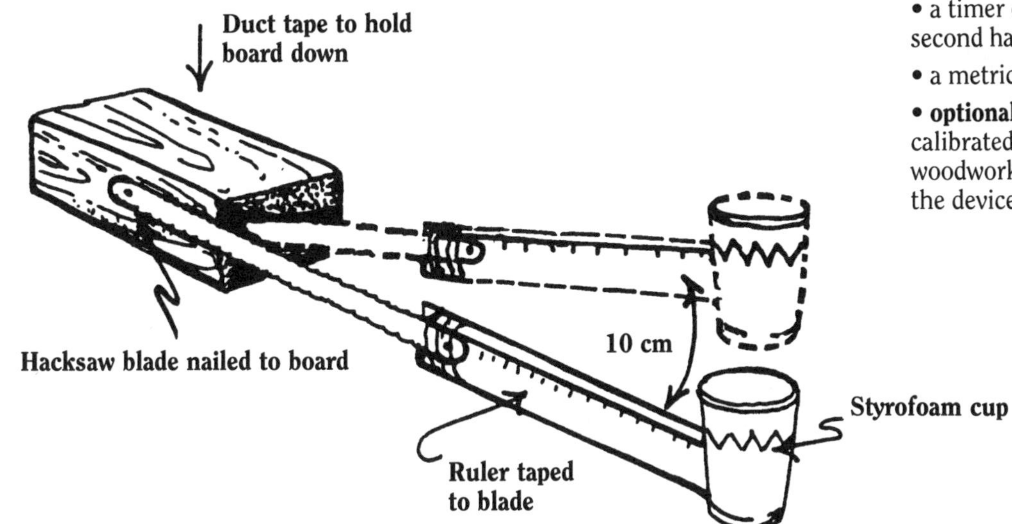

Materials

Each group will need

- a 2 x 4 board about 46 cm long
- 3 nails
- a hacksaw blade
- a ruler or wood molding about 46 cm long
- a bolt, washer, and nut
- a styrofoam cup
- duct tape
- 3 fishing sinkers (weighing approximately 50 g each)
- a timer or clock with a second hand
- a metric ruler
- **optional:** a spring scale calibrated in newtons and a woodworker's C-clamp to hold the device steady

◆Procedure

1. Assemble the mass-measuring device shown in the diagram.

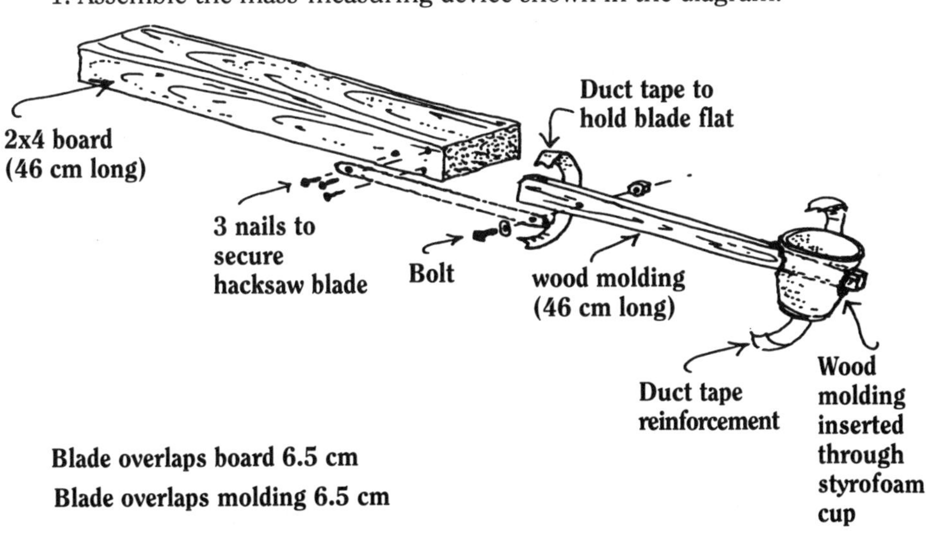

! Place a layer of duct tape over the cutting edge of the saw blade to make the device safer for student use.

! Make certain the board is securely attached to the table surface with either a clamp or duct tape.

Place the mass-measuring device on the edge of a table so that the blade, ruler, and cup part of the device (henceforth called the "arm") can swing parallel to the floor. The arm should not hit anything as it goes back and forth.

2. Place one finger in the cup and pull the arm back 5 cm from its rest position. When you pull the arm back, you exert a **force** on the arm. You can think of a force as a push or a pull in a particular direction. Note how hard you have to pull compared to the force required to pull it back 10 cm; then pull it 15 cm from the rest position.

State in your own words how the amount of force required to move the arm changes as the cup is pulled farther and farther from the rest position.

____

____

____

Optional: If you have a spring scale, attach it to the cup and measure the force required to pull the arm 5 cm, 10 cm, and 15 cm. Note your scale readings below:

____

____

____

Vocabulary

- **Force:** A push or pull in a particular direction that can be applied to an object.
- **Inertial balance:** A device used to measure mass by making use of the relationship between mass and inertia.
- **Newton:** A unit used in measuring force. One newton is the amount of force required to accelerate a 1-kg mass 1 m/s^2.

3. Pull the cup back 10 cm from the rest position and let go. The arm should swing smoothly back and forth at least 15 s. The longer the arm swings, the better. If it does not swing evenly, check the following:

- Is the saw blade firmly attached to the 2 x 4 support?
- Is the support held down firmly with duct tape or a C-clamp?
- Are the arm and cup firmly taped to the blade?
- Does the arm swing *parallel* to the floor, *not* up and down?

Measure how long it takes the cup to stop swinging.
Time = ____ s.

4. Start the arm swinging again by pulling it back 10 cm from the rest position. Measure how long it takes the arm to swing back and forth 20 times (in seconds). Repeat your measurement two more times, and enter your observations in the data table.

Data Table

Time required for arm to swing 20 times

	Trial 1	Trial 2	Trial 3	Average time
empty cup	_____	_____	_____	_____ s
cup + 1 sinker	_____	_____	_____	_____ s
cup + 2 sinkers	_____	_____	_____	_____ s
cup + 3 sinkers	_____	_____	_____	_____ s

5. Place 1 sinker in the cup. Start the arm swinging again by pulling it back 10 cm from the rest position. Measure how long it takes to complete 20 swings. If the sinker rattles as the arm is swinging, wedge it in the cup or use a small piece of duct tape to hold it in place.

6. Make the rest of the observations called for in the data table. Place the specified number of sinkers in the cup. Time 20 swings of the cup, and repeat this measurement three times for each set of sinkers. Compute the average time required for each set of sinkers.

7. In your own words, describe how the time required for 20 swings changes as more mass is added to the cup.

____

____

____

____

8. What part of the device gives the cup and sinkers being measured a push (exerts a force)?

____

9. How was the force kept constant for each trial?

____

____

____

____

____

GUIDE TO ACTIVITY 7

Defining Mass as "Difficult-to-Moveness"

◆What is happening?

The results of this activity can be explained by the fact that as an object's mass increases, so does its **inertia**, or "difficult-to-moveness." As more sinkers are placed in the cup, the amount of mass being moved by the arm increases. As the mass increases, the inertia ("difficult-to-moveness") of the cup also increases, and the arm takes longer to complete 20 swings.

The mass-measuring device used in this activity is a simple inertial balance. Its spring (the saw blade) applies the same amount of force to the cup for each trial. The arm swings more slowly as sinkers are added to the cup because the *force* being applied to the cup remains the same, but its *inertia* ("difficult-to-moveness") increases with each additional sinker.

The initial amount of force applied by the device depends on the point of release of the arm. A larger displacement of the arm before it is released produces a greater initial force acting on the mass in the cup. For this reason, it is essential to release the arm from the same position on each trial. Otherwise, the results for each trial will vary, and interpreting the observations will be difficult.

The arm of an inertial balance changes speed (accelerates) more slowly when mass is added to the cup. Loading the cup increases the resistance to motion (inertia or mass) of the balance, but the push (force) being exerted by the arm will be about the same for each trial if the point of release is kept constant. Therefore, the balance takes longer to complete 20 swings when mass is added to the cup.

Inertial balances will work in a weightless environment such as an orbiting spacecraft, since they measure the property of "difficult-to-moveness." Pan balances or spring scales depend on gravity to exert a force on the mass being measured. They cannot be used in spacecraft.

Comparing pendula and inertial balances

A question that may occur to students using inertial balances is "Why does adding mass to the inertial balance slow its rate of swinging, while changing the mass of a pendulum does not affect the pendulum's period?" Since swinging back and forth occurs with both devices, it does seem reasonable to expect both to behave in a similar fashion. The solution to the seeming contradiction depends on the amount of force acting on the mass and the direction of the motion of the mass relative to the direction of the force.

The addition of mass to the pendulum did not slow down the swinging rate because the added mass also resulted in a greater gravitational pull (objects with more mass are heavier). The added mass made the pendulum harder to get going or more resistant to change in motion, but there was a greater downward gravitational pull to overcome that greater resistance. The greater "difficult-to-moveness" was offset by a greater force so the rate of pendulum swing remained the same.

The inertial balance slowed down when more mass was added to the cup because the sideways force exerted on the cup by the hacksaw blade did not increase when more mass was added to the cup. Here the greater "difficult-to-moveness" was added to the cup without an increase in the force causing the back and forth movement. With greater "difficult-to-moveness" and the same sideways force, the rate of back and forth motion was reduced. Just as with the pendulum, when the mass increased, the gravitational pull increased. This greater pull did not result in greater motion because the pull was downward in direction and the motion was sideways in direction. The increased downward gravitational pull on the weights was matched by an increased upward push on the weights by the cup.

Applying the second law of motion to an inertial balance

The results of this activity can be explained in a more mathematical fashion by using Newton's second law of motion, which is stated F = ma (force = mass X acceleration). In the special case of our inertial balance:

force = push applied by the saw blade; remains the same

mass = amount of matter put in the cup (plus the mass of the cup and the molding or ruler)

acceleration = the rate at which the speed of the mass changes. This is *inferred* from the amount of time required to complete 20 swings. Acceleration is not measured directly. A short time to complete 20 swings implies a large acceleration; a long time to complete 20 swings implies a small acceleration.

In this activity, the initial force (F) exerted by the saw blade is kept the same for each trial by pulling back the blade the same amount each time. Therefore, in this special case the equation F = ma may be simplified:

$$F = \text{constant} = m \times a$$

Since m X a = constant, the inertial mass can be defined as:

$$m = \text{constant} / a$$

Examining this equation shows that the only way the relationship can remain numerically balanced when m is increased (more matter is added to the cup) is for a to decrease (it takes longer to complete 20 swings). In other words, the equation suggests that more massive objects are harder to set in motion than less massive objects.

In fact, this is exactly the result observed experimentally. When the cup is heavily loaded, it swings more slowly. Slower swinging means that *decreased* acceleration is taking place. Since the saw blade is pulled back the same distance each time, the initial force is constant.

◆Time management

If the inertial balances are constructed before class, one class period (40–60 minutes) should be enough time to complete the activity and discuss the results.

◆Preparation

You may wish to ask a parent or older student who is an experienced woodworker to assemble a set of inertial balances for you.

The most reproducible results will be obtained when the weights do not rattle inside the cup. If the weights move, wedge them in the bottom of the cup, or use a small piece of masking tape to hold them down. The small amount of extra mass will not measurably affect the results.

The length of the 2 x 4 board used in the balance is not critical. Short lengths of boards suitable for building these balances can often be obtained free from carpenters working on construction sites.

The **newton** (N) is the internationally recognized unit of force. Whenever possible, encourage students to measure force in newtons. If spring scales calibrated in newtons are not available, spring scales labeled in grams can be used. It should be emphasized that grams are a measure of *mass* rather than *force*. However, a spring scale will indicate 100 g when a force of 1 N is applied to it. Therefore, if you are using a scale labeled in grams, you can divide the reading in grams by 100 to obtain newtons.

Example: Reading on scale = 125 g

125 g ÷ 100 = 1.25 N

Any of the following may be used for testing the balance:

- 2-ounce lead fishing weights (These are inexpensive, and have a mass of about 50 g each. The pyramid style sinker has the advantage that its flat sides make it unlikely to move inside the cup.)
- 50-g laboratory weights
- pieces of gravel
- common nails (preferably 16 penny or larger size nails)
- any other solid object with a mass that can be increased in increments of 100–150 grams and that can be fitted securely in the cup

◆Suggestions for further study

Place two small boxes on a table. Have students try to determine, without lifting the boxes, which of the two would feel heavier when lifted. Students are allowed to touch and move the boxes, but they are not allowed to lift them. If the boxes are shaken back and forth without lifting, one will feel harder to move back and forth. That object has greater "difficult-to-moveness" and so has more mass. Near the Earth, objects with greater mass also have greater weight. When the boxes are lifted, weight is being experienced.

Ask students what they could do to get the same time for twenty swings for two different masses placed in the cup of the inertial balance. To get the same acceleration or rate of change of motion, you would have to add a greater force to the object with greater mass or "difficult-to-moveness." Using stiffer or more hacksaw blades would increase the force on the cup. It may be possible to get a greater force by pulling the hacksaw blade back farther before release.

◆Answers

2. You have to pull harder to move the arm 15 cm than to move it 10 cm or 5 cm; the farther you move it, the harder you have to pull.

Sample data

Average time required for 20 swings:

empty cup	10.6 s
cup + 1 sinker	21.7 s
cup + 2 sinkers	30.4 s
cup + 3 sinkers	44.5 s

Note: These results were obtained using a balance built to the dimensions given in the activity and using 2-ounce fishing sinkers for weights. *Results will vary from balance to balance,* depending on its dimensions, stiffness of the blade, length of the arm, etc.

7. As the mass in the cup is increased, the time for 20 swings increases.

8. The hacksaw blade provides the force (the push) to move the masses.

9. By pulling the arm back exactly 10 cm and by releasing it smoothly, the force exerted by the arm on the masses in the cup will remain constant for each trial.

ACTIVITY 8 WORKSHEET

Applying Force to a Rubber Band

◆Background

The weight and mass of an object are not the same. **Weight** is a force; it occurs because of the gravitational attraction between an object and the Earth. An object's mass is a measure of how much of the object there is —how much matter makes up the object. The more mass in an object, the greater the weight. In this activity, you will be able to see how this downward force, weight, increases as mass increases.

◆Objective

To observe how the force on a rubber band increases as the mass attached to it increases

◆Procedure

1. Hook both paper clips on the rubber band. Hang the rubber band on the ring stand by attaching one of the paper clips to the clamp. (You may need to bend the paper clip some, depending on the diameter of the clamp.) Slide the other paper clip to the lowest point of the rubber band. Bend the bottom clip into a hook (see diagram below) so that you will be able to hang weights on it.

2. Measure the distance between the points labeled A and B in the diagram. Points A and B are where the paper clips touch the rubber band. This distance will be called the *initial length*. Determine the initial length of the rubber band to the nearest 0.1 cm. Record the initial length in the data table.

3. Hang sinkers on the bottom clip one at a time. Measure the distance between A and B to the nearest 0.1 cm each time that you attach a sinker. In the data table, record the total amount of mass hanging from the rubber band every time you measure its length (2-ounce fishing sinkers have a mass of approximately 50 g each).

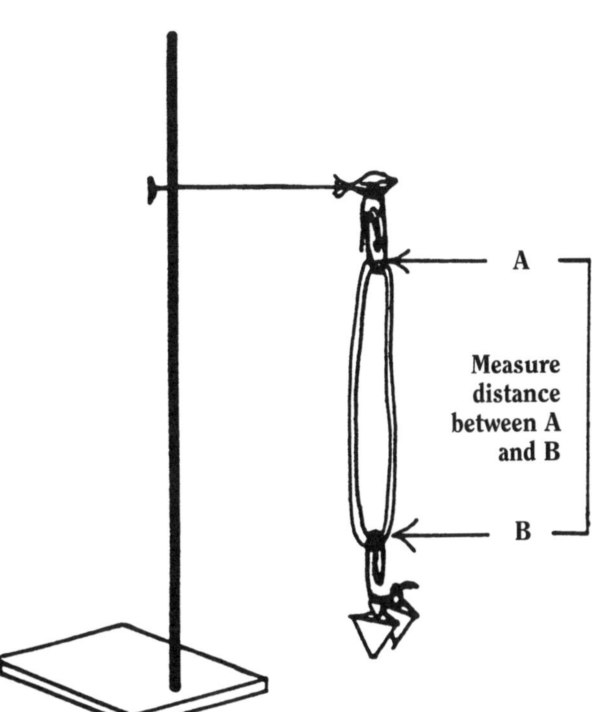

Materials

Each group will need

- safety goggles
- a heavy-duty rubber band
- 6 50-g weights (use either lead pyramid fishing sinkers or similar-size laboratory weights)
- a ring stand with a clamp attached near the top
- 2 paper clips
- a meter stick or ruler
- graph paper
- **optional:** string for tying a support loop to each weight

Vocabulary

- **Weight:** A force due to the gravitational attraction between an object and the Earth. It is a downward force that acts on the object.

! Safety goggles are strongly recommended both for setting up and performing this activity.

! Do not break the rubber band! If the rubber band seems to stop stretching, do not attach additional sinkers.

Data table

The stretch of a rubber band

Total mass hanging on bottom paper clip	Length of rubber band in centimeters (the distance between A and B)
0 g	_______ cm (initial length)
_______ g	_______ cm
_______ g	_______ cm
_______ g	_______ cm
_______ g	_______ cm
_______ g	_______ cm
_______ g	_______ cm

4. Describe in your own words how the length of a rubber band changes as sinkers are hung from it.

____

____

____

5. A **force** is a push or pull in a particular direction. Each sinker hanging from the hook exerts a *downward* force on the rubber band. Complete the following sentence relating the force acting on the rubber band with the band's change in length:

As the force acting on the rubber band _______________, the length of the rubber band _______________.

6. The rubber band exerts a force on the sinkers hanging from it. In what *direction* does the rubber band exert force—up or down? _____________

7. Predict what would happen if 25 additional sinkers were hung from the rubber band.

____

____

____

Length of rubber band (cm)

50
45
40
35
30
25
20
15
10
5
0

100 200 300 400 500 600

Total mass (g)

8. Use the observations you recorded in your data table to complete a graph similar to the one shown.

Label the x-axis (horizontal axis) "Total mass (grams)," and the y-axis (vertical axis) "Length of rubber band (centimeters)."

Do the points on the graph follow any particular pattern?

____

____

____

GUIDE TO ACTIVITY 8

Applying Force to a Rubber Band

◆What is happening?

In this activity, you measure how much a rubber band stretches when lead fishing sinkers are attached to it. Attaching a sinker exerts a downward force on the rubber band; the rubber band exerts an upward force on the sinker. The rubber band stops stretching and the sinker comes to rest when the the upward force exerted by the rubber band is equal to the sinker's downward force.

Weight is a downward force due to the gravitational attraction between an object and the Earth. The weight of the sinker determines the distance that the rubber band stretches. The rubber band will stretch more if additional sinkers are attached to it, because the additional sinkers increase the downward force acting on the rubber band.

The increase in length of the rubber band is proportional to the amount of force being exerted by the sinkers. Graphing its length shows that the rubber band's length changes in a regular, predictable fashion as weight is added.

Spring scales used for measuring forces take advantage of a similar pattern of stretching that occurs in metal springs. As force is applied to a spring, it stretches in a regular pattern similar to that of the rubber band used in this activity. By attaching a pointer to the spring in a scale, the distance that the spring stretches can be converted into an indication of the force (in newtons) acting on the spring.

The more mass an object has, the greater will be its weight near the surface of the Earth. Weight and mass are *not* the same thing, however. *Mass* is the quantity of matter. An object's mass is determined by its total number of protons, neutrons, and electrons—the subatomic particles it possesses. Subatomic particles are particles of matter smaller than an atom. Every object is the result of some combination of these basic particles. Since subatomic particles have mass, the more particles there are in an object, the more mass the object has. Activities 6 and 7 showed that you can also describe an object's mass in terms of its "difficult-to-moveness," or inertia.

◆Time management

One class period (40–60 minutes) should be enough time to complete the activity and discuss the results.

◆Preparation

It is a good idea to test the rubber bands you will use with the weights you plan to use before assigning this activity. Rubber bands vary considerably from batch to batch. You may find it necessary to use different weights, or to modify the procedure in one of the ways described below.

The best rubber bands for this activity are the natural rubber type that (unstretched) are about 0.4 cm wide and 9 cm long (in other words, about 18 cm in circumference).

If the rubber band will not stretch enough for accurate distance measurements, try cutting it in half and tying the paper clips to each end. This should increase the distance it stretches as each weight is added.

If your rubber bands are too small and stretch to their maximum length too quickly, attach two or three bands between the paper clips before adding weights.

A 2-ounce lead fishing sinker has a mass of about 50 g. These are cheap, readily available, and can be freely substituted for the calibrated lab weights. The small variations in the fishing weights will not affect the results measurably. Tying a short loop of string to each weight may make it easier to hang the weights from the paper clip.

This activity can also be done without purchasing any weights: hang an empty can from a smaller rubber band by punching holes at the open end and making a paper clip handle (like the handle on a bucket). Pour water into the can in 10-ml increments (10 ml of water has a mass of 10 g), and measure the stretch of the rubber band. This set-up can be hung from a nail if no ring stands are available.

◆Suggestions for further study

Does buoyancy affect the total pull the weights exert? Measure the stretch while the weights are submerged in water.

Modify the rubber band apparatus so that it can be used to weigh unknown weights in the manner of a spring scale.

◆Answers

Answers will vary depending on the amount of weight added and the strength of the rubber band. The following data set is included *only as an example*; student answers will be different from those given.

Sample data

The stretch of a rubber band

Total mass hanging on bottom paper clip	Length of rubber band in centimeters (the distance between A and B)
0 g	7.4 cm (initial length)
50 g	8.2 cm
100 g	9.4 cm
150 g	11.3 cm
200 g	13.2 cm
250 g	15.1 cm
300 g	16.9 cm

4.The length of the rubber band increases as additional weights are hung from it.

5. As the force acting on the rubber band <u>increases</u>, the length of the rubber band <u>increases</u>.

or,

As the force acting on the rubber band <u>decreases</u>, the length of the rubber band <u>decreases</u>.

6. The rubber band exerts an *upward* force on the sinkers (the sinkers exert a downward force on the rubber band).

7. The rubber band would break because the downward force of the sinkers would be greater than the maximum upward force that the rubber band can exert.

8. The data should produce a graph that is approximately linear. There may be some variation from a linear relationship for small amounts of weight and large amounts of weight. If the rubber band is overstretched, the graph will flatten out as additional weights are added.

MODULE 3

Constant Speed Versus Acceleration

◆Introduction

• How fast is a communications satellite moving?

• How quickly can a race car reach its top speed?

• How far did a golf ball travel?

In order to answer the above questions about moving objects, a physicist would want to know several facts:

• How far did the object move while it was being observed?

• How much time passed while the object's motion was observed?

• In what direction did the object move?

• What changes in rate or direction of motion were observed?

In this module, information about "How far?" and "How much time?" are combined to produce a quantitative description of motion: **speed**. Adding information about "What direction?" for a moving object produces a more specific quantitative description of motion: **velocity**. Measurements of changes of speed and velocity are used to determine **acceleration.**

◆Instructional Objectives

After completing the activities and readings for Module 3, you should be able to

- measure the displacement of a moving object [Activities 10, 11, and 12]
- employ Δ notation to represent displacement and intervals of time [Activity 10 and Reading 7]
- calculate the speed of a moving object [Activity 10]
- determine whether or not an object is moving at a constant speed [Activity 11]
- explain the differences between speed, velocity, and acceleration [Activities 9–12 and Readings 7 and 8]
- define acceleration [Reading 8]
- demonstrate at least three ways of determining whether or not acceleration is occurring [Activities 12, 13, and 14]
- identify forces that can produce acceleration [Readings 8 and 9]
- give an example of an operational definition [Activities 10–13 and Reading 8]

◆Preparation

Study the following readings for Module 3:

Reading 7: The Magnitude of Motion: Speed

Reading 8: An Intuitive Approach to Defining Acceleration

Reading 9: Algebraic Representation of Acceleration

◆Activities

This module includes the following activities:

Activity 9: *Eureka!*—#5: Acceleration part II

Activity 10: Determining the Speed of a Toy Car

Activity 11: Is the Speed of a Toy Car Constant?

Activity 12: Determining the Acceleration of a Toy Car

Activity 13: Using an Accelerometer to Test for Changes in Speed

Activity 14: Using "Tappers" to Investigate the Fall of a Thread

ACTIVITY 9: VIDEOTAPE

Eureka! #5—Acceleration part II

◆Background

Program 5 of *Eureka!* defines and illustrates specific methods of measuring the phenomena of **speed**, **constant speed**, and **acceleration**. This segment can serve as an excellent advance organizer for the study of speed and acceleration. (Even though this segment is labeled as "part II" of the programs on acceleration, it requires no advance knowledge of the concepts of speed and acceleration. You will have the opportunity to view programs 3 and 4 in the next module.)

◆Time management

The running time of the videotape is 5 minutes. At least 15 minutes should be allotted to introduce, run, and discuss the videotape. You may wish to play the videotape at the end of a lesson to reinforce the concepts presented. This *Eureka!* program contains a great deal of information. You may wish to show it more than once. It is particularly useful for summarizing the concepts presented in Activities 10, 11, and 12.

◆Comments on the videotape

The videotape starts by asking, "How can we measure acceleration?" Most people know that acceleration has something to do with speed and time (an illustration of a clock and a speedometer calibrated in kilometers per hour [km/hr] are shown on the videotape). However, they are not sure exactly what is meant when a train's acceleration is described as 1 meter per second squared; they may also be uncertain about the units of speed. Therefore, a logical first step toward understanding numerical representations of speed and acceleration is knowing their units, and being able to convert the units to the most useful terms. The top speed of the train used for the illustration, 36 km/hr, is shown to be mathematically equivalent to 10 meters per second (m/s). These calculations are shown in detail in the film.

After dealing with the numerical representation of speed, a graphic representation of constant speed is shown: a constant top speed of 10 m/s means that the train traverses 10 m of track each second. In the film, the distance traveled each second is marked by the engineer of the train, who drops a rock behind the locomotive with every tick of the clock. The rocks land at equal 10-m intervals down the track.

Having established what is meant by constant speed, the program then describes acceleration in terms of the motion of the train: "Acceleration means that with every second that passes, the train advances an *increasing* number of meters until it is eventually moving at a speed of 10 m/s...but it didn't start out that way"* (it didn't start out moving at 10 m/s, because the initial speed was 0). The distance traveled by the train in every one-second interval is again illustrated by having the engineer drop rocks off the back as it goes from a standstill (0 m/s) to its top speed (10 m/s) in 10 s. This concept of acceleration is used as the basis for the activity "Determining the Acceleration of a Toy Car" included in this module.

The program ends with a brief statement that acceleration is measured in meters per second squared (m/s^2). (A more complete discussion of the algebraic derivation of this unit is included in Reading 9.)

Concept summary

"Since acceleration refers to rate of change of speed, and since speed can be measured in meters per second, acceleration is measured in meters per second per second ."*

Vocabulary

- **Acceleration:** The rate at which an object speeds up or slows down.
- **Constant speed:** An object is traveling at constant speed if it moves the same distance within every equal time interval.
- **Speed:** The rate of motion; speed combines information about how far an object travels with how long it takes to travel that distance.

**Eureka!* Produced by TVOntario © 1981.

ACTIVITY 10 WORKSHEET

Determining the Speed of a Toy Car

Materials

Each group will need

- a meter stick
- a battery-operated electric model car or truck
- a timer or clock with a second hand
- 15 cm of masking tape

◆Background

The speed of an object is determined by the amount of time it takes an object to move over a particular distance. If we want to know how fast something is going we only have to measure the distance it travels and the time it takes to travel it. We would then have all the information we need to determine its speed.

◆Objective

To measure the speed of a toy car

◆Procedure

1. Measure a 1.5-m track for your car on a smooth surface, such as a table top or floor. Use masking tape to mark "start," "time," and "finish" lines like those shown below.

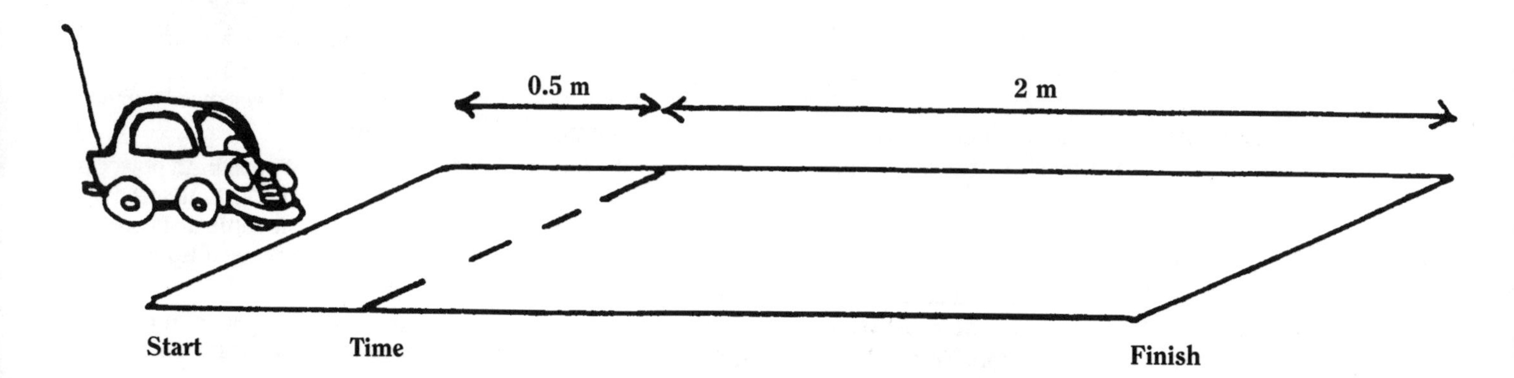

2. Switch on your car and place it on the start line. When it reaches the time line, begin timing how long it takes to reach the finish line. Record your time observation in the data table as "Trial 1." Time three runs using this procedure and record your observations.

Data table

	Trial 1	Trial 2	Trial 3
Seconds for car to go 1 m	_____	_____	_____

3. Compute the average for Trials 1–3.
   Average time interval to travel 1 m = _______ s.

4. The relationship between speed and distance is stated:

$$\text{Speed} = \frac{\text{distance moved}}{\text{time interval}}$$

While it is being timed for this activity, the car moves a distance of 1 m between the time and finish lines. Dividing this distance (1 m) by the average time interval that you computed (for step 3) gives the average speed of the car for the three runs.

Substitute your value for the average time interval in the equation below, and calculate the average speed of the car.

$$\text{Average speed} = \frac{\text{distance moved}}{\text{average time interval}} = \frac{1\ \text{m}}{\_\_\_\_\_\_\ \text{s}} = \_\_\_\_\_\_\ \text{m/s}$$

5. Why is it best to repeat your measurements several times in an experiment like this one?

____

____

6. What is the purpose of letting the car run 0.1 m before starting to measure the time?

____

____

7. If the car that you are using has different speed settings, change the speed setting and repeat the measurements. How does the calculated value for the car's speed change?

____

____

8. Predict how many seconds it will take your car to travel 150 cm.
Predicted time = __________.
Test your prediction.
Actual time taken to travel 150 cm = __________.

9. Predict how many seconds it will take your car to travel 75 cm.
Predicted time = __________.
Test your prediction.
Actual time taken to travel 75 cm = __________.

10. Pick a time for your car to travel. Predict a distance your car will travel in that time. Test your prediction. Compare your prediction and the actual distance covered.

GUIDE TO ACTIVITY 10

Determining the Speed of a Toy Car

◆What is happening?

The measurement of the amount of time required for the toy car to move over a known distance (1 m) allows a direct calculation of the speed of the car. The measurements are made using meters for distance and seconds for time. Combining them gives the preferred laboratory units of speed, meters per second (m/s).

This activity gives a simple example of a general way of calculating speed by determining an object's **displacement** within a time interval. Displacement is simply the distance that an object moves. A way of stating the relationship between speed, displacement, and the time during which the motion is measured is:

speed = displacement / time interval

When calculating displacement, the positions of objects moving in a straight line are often treated numerically as though the object were moving along the x-axis of a graph. Each position of the object is shown as a value of x. This allows the equation for speed to be stated:

speed = the change in x / time interval

Scientists frequently use the symbol Δ **(delta)**, to symbolize "*the change in* ___" or "*calculate an interval of* __." Using delta notation, the equation for speed can be stated:

speed = $\Delta x / \Delta t$

Additional information about the use of Δ notation and calculations of speed is given in Reading 7. You may wish to work through the sample speed problems given in that reading after completing the activities.

Vocabulary

- **Delta notation:** A symbolic indication of an interval using the Greek letter delta, written Δ; for any quantity x, Δ x means *"the change in x"* or *"calculate an interval of x."*
- **Displacement:** The distance that an object moves.

◆Time management

One class period (40–60 minutes) should be enough time to complete the activity and discuss the results.

◆Preparation

These measurements can be performed using any battery-powered model car that runs at a fairly constant speed. Your students (or their younger siblings) can probably provide a selection of cars that can be used for this activity. In addition to saving departmental equipment funds, having students provide the vehicles used in the activity is a good motivational technique.

If you choose to buy electric cars for this activity, the Stomper™ brand of car is a good choice. They are inexpensive, relatively rugged, and use only one battery. Some of the Stomper cars have additional features that are useful for later activities: a "freewheeling" (non-motor powered) setting, and "high" and "low" speed settings.

Some cars have slightly uneven wheels that cause them to curve while running down the length of the 1.5-m course. If the curve is small and relatively consistent from run to run, you can still use the data collected to calculate speed. Lay a string along the approximate curving path the car follows. Measure the length of the string to estimate the actual distance the car travels during the run. Use this distance rather than 1 m to calculate the speed.

If the curving path seems to be causing variations in the times recorded, shorten the course. One suggested layout: allow at least 0.25 m from the start to the time line so that the car can reach a constant speed. Place the finish line 1 m from the time line. Modify the instructions for calculations accordingly.

◆Suggestions for further study

Have students use their cars and data to make predictions about how long it would take to travel different distances. For example, if the car went 1 m in 4 s, how long would it take to go 4 m? How long would it take to go 3 m? Have students test their predictions. Ask students to use their data to predict how far their car would go in different time intervals. For example, if the car travels at an average speed of 0.25 m/s, how far would it go in 10 s?

◆Answers

The exact amount of time required to complete the 1.5-m course *will vary from car to car.* The three runs for a single car should be very similar; if the times vary greatly, repeat the measurements. The average for the three runs will probably be between 3 s and 8 s.

3. Using a Stomper, we obtained an average time of 4 s.

4. Speed = 1 m / 4 s = 0.25 m/s. *This answer will be different for different cars!*

5. It is hard to measure the time exactly; the speed of the car might change; repeating gives a better basis for comparing results.

6. The car needs to run a short distance to reach a constant speed.

7. Answers will vary depending on the car/battery combination used.

ACTIVITY 11 WORKSHEET

Is the Speed of a Toy Car Constant?

Materials

Each group will need

- a meter stick
- a battery-operated electric model car or truck
- a timer or clock with a second hand
- 15 cm of masking tape
- a "mini drip timer" (the push-pull top from a bottle of dishwashing liquid detergent)
- 100 ml of water with food coloring added
- a medicine dropper
- paper towels or cloths for wiping up water droplets

Hint: *The timer drips more slowly if it is low on water. This can lead to erroneous inferences about the car's speed. Keep the timer as full as possible, and use only the first 8 to 10 drops for analysis of the car's speed. If the timer is obviously dripping slower at the end of the run, only measure the spots of water numbered 3–8. This will give 5 reliable data points.*

◆Background

How can we tell if something is traveling at a constant speed? If an object travels the same distance during each successive time interval, then an object is traveling at a constant speed. In this activity, you'll be able to determine how far a toy car moves in each time interval so that you can answer the question, "Is the car traveling at a constant speed?"

◆Objective

To determine if a toy car is traveling at a constant speed

◆Procedure

Part I Calibrating the mini drip timer

1. Hold the top from a bottle of dishwashing detergent upside down over the container of colored water. Use the medicine dropper to put the colored water into the bottle top.

2. Open the top until water drops out at a rapid, steady rate. Aim for between 2 and 5 drops per second. If the water comes out in a stream or drops slowly and irregularly, adjust the sliding closure down or up a little. Continue adjusting the opening of the top until you can get fast, even dropping rates. (**Note:** This process involves trial and error. Be systematic and try to move the closure only a tiny distance each time you adjust it. Once you get fast, regularly spaced drops, be careful not to move the sliding closure.)

3. After the timer is dripping at a fast, regular rate, see how long it takes for it to drip 10 times. Fill in the following data table:

4. Calculate the average time for the three trials.

Average time per 10-drop trial
= ________ s

Data table

Trial	Seconds for 10 drops
1	_____ s
2	_____ s
3	_____ s

5. Calculate the average number of drops per second. (Divide the average time per trial into the number of drops counted per trial.)

10 drops/(average time per 10-drop trial)
= ______ drops per second

Part II Using the mini drip timer to mark distances traveled by the car

6. Tape the timer to the car so that its drops of water will land clear of the path of the wheels. *Be careful not to change the setting of the tip*, or recalibrating the drip rate will be necessary!

7. Place the car and timer over the starting position of the 2.5-m course and fill the timer with enough colored water to produce about 15 steady drops. Hold your finger over the opening of the timer so that it does not leak. Be careful not to change the setting of the tip!

8. Switch on the car, and let it run down the course. Observe the pattern of the drops of colored water on the table. Are they the same distance apart? Describe the pattern of drops.

____

____

____

9. Measure the distance between the drops to the nearest 0.1 cm, beginning with the *third drop*. (Starting with the third drop allows the car to reach a constant speed before you begin your measurements.) Measure from the center of one drop to the center of the next drop (see diagram below).

Measure distances between centers of drops

Time **Finish**

10. Record your measurements in the following data table. (**Note:** You may only get 5 or 6 good clear drops past the third drop. That is sufficient for this activity. Record only those drops you can measure accurately. If the drip timer is dripping unevenly, recalibrate it and repeat the run.)

11. After you record the measurements for Trial 1, dry off the course, refill the drip timer, and repeat the procedure. Record your observations for Trial 2 and Trial 3 in the data table.

12. If time permits and your car has more than one speed, change the speed and repeat the measurements. Record your observations on a separate sheet of paper.

13. Does the car travel the same distance during each time interval between the drops? How can you tell?

____

____

____

____

____

Data table

Drops	**Distance between drops**		
	Trial 1	**Trial 2**	**Trial 3**
1 & 2		do not measure	
2 & 3		do not measure	
3 & 4	____.__cm	____.__cm	____.__cm
4 & 5	____.__cm	____.__cm	____.__cm
5 & 6	____.__cm	____.__cm	____.__cm
6 & 7	____.__cm	____.__cm	____.__cm
7 & 8	____.__cm	____.__cm	____.__cm
8 & 9	____.__cm	____.__cm	____.__cm
9 &10	____.__cm	____.__cm	____.__cm

GUIDE TO ACTIVITY 11

Is the Speed of a Toy Car Constant?

◆What is happening?

This is a real life demonstration of the hypothetical example (shown in *Eureka!* #5) of dropping rocks off the back of a moving train once per second. The drip timer (if it is working correctly) releases a drop of water at a constant rate. The drop of water marks the location of the toy car at regular time intervals.

The distance between the drops shows how far the car traveled during the time between drops. (The actual time interval between drops will vary depending on the timer being used, but should be about 1/3 s to 1/2 s.) Assuming that the apparatus is working correctly, all the distances between the drops should be approximately the same after the car reaches a constant speed. This result demonstrates an **operational definition** of constant speed: an object moving the same distance during every equal time interval is moving at a constant speed.

Vocabulary

- **Operational definition:** A non-mathematical definition of a term that includes an operation to perform and observations to make in order to determine whether the defined phenomenon is present.

◆Time management

One class period (40–60 minutes) should be enough time to complete the activity and discuss the results.

◆Preparation

Calibrating the mini drip timer is probably the most difficult part of this activity. You may find that some brands of detergent have tops that work better than others.

If your students have not already done Activity 3, "Building a Drip Timer," you may wish to have them do it before attempting to work with the mini drip timer. For information about the cars used in this activity, refer to the "Preparation" section of Activity 7.

Some cars have slightly uneven wheels that cause them to curve while running down the length of the course. If the curve is small and relatively consistent from run to run, you can still use the data collected to check for constant speed. Simply ignore the curve, and measure the distance from drop to drop as though the car had traveled in a straight line.

◆Suggestions for further study

Slow the car down by taping a lead weight to it. What do you predict about the distance between the drops for a slower car?

A more accurate (but more complicated and expensive) method for demonstrating constant speed is to make multiple strobe exposures of a moving object. The resulting photograph will show how far the object moved between flashes.

◆Answers

Keep adjusting the mini drip timer until you get about 3 drops per second. *The exact rate will vary from group to group.* The answers given below are accurate only for *one specific dropper!*

3. Using a Palmolive™ top, 10 drops fell in 3.1 s for each trial.

4. 3.1 s

5. 10 drops ÷ 3.1 s = 3.2 drops per second

8. The drops are very evenly spaced after the car reaches a constant speed (after it has traveled about 10–20 cm from the start line).

9–11. Our car traveled an *average* of 9 cm between centers of drops. The actual measurements varied between 8.5 cm and 9.5 cm.

13. Yes—the car travels a constant distance between every pair of drops. You can infer that the distance that the car travels during the time interval between drops is the distance measured between each pair of drops. Since the drops occur at a steady rate (3.2 drops per second for the test car), the car must be moving at a constant speed since it moves the same distance during every equal time interval.

ACTIVITY 12 WORKSHEET

Determining the Acceleration of a Toy Car

Materials

Each group will need

- a meter stick
- a model car
- a "mini drip timer" (the push-pull top from a bottle of dishwashing liquid detergent)
- 100 ml of water with food coloring added
- a medicine dropper
- a watch or clock with a second hand
- a table with a smooth surface about 2 m long
- bricks or old books for propping up one end of the table
- 15 cm of masking tape
- towels or cloths for wiping up water droplets

◆Background

You can guess whether an object is accelerating by watching it. In this activity, you will use a more objective, and more accurate, method to determine whether or not an object is accelerating—making time and distance measurements for an object. You can then use your results to see how reliable your guess was.

◆Objective

To determine if a toy car is accelerating

◆Procedure

Part I Observing acceleration down a ramp

1. Place the car on the table top or on a long, wide board. (If a board is used, it should be approximately 2 m long and 0.4 m wide.) Tilt the table until the car *just begins to roll*. Prop up the table legs by putting bricks or old books under them.

2. Place the car at the high end of the incline and release it. *Do not push it*. Does it seem to change speeds as it moves down the ramp? Describe its motion.

___

___

___

Part II Studying acceleration by combining measures of time and distance traveled

Hint: *A 10-cm or lower incline should be sufficient. You should keep the incline as shallow as possible. You want to establish a steady acceleration—not try for a new land speed record. High speeds may hinder the analysis of results later.*

3. Set up a mini drip timer so that it releases about 2 or 3 uniform drops per second.

4. Tape the timer to the car so that its drops of water will land clear of the path of the wheels. *Be careful not to change the setting of the tip*, or recalibrating the drip rate will be necessary!

5. Place the car and timer on the top of the incline and fill the timer with enough colored water to produce 10 to 15 steady drops.

6. Hold your finger over the opening of the timer so that it does not leak. *Be careful not to change the setting of the tip!*

Release the car and let it run down the incline. Observe the pattern of the drops of colored water on the incline. Are they the same distance apart? Describe the pattern of drops in your own words.

___

___

Part III Using the mini drip timer to record the distance a car travels per unit of time

7. Measure the distance between the drops on the incline to the nearest 0.1 cm. Measure from the center of one drop to the center of the next drop (see diagram below).

8. Record your measurements in the data table. (**Note:** You may only get 5 or 6 good clear drops. Record only the drops you can measure accurately. There may be a puddle near the starting point since it is difficult to release the car without dropping some extra water. For this reason, you may wish to start your measurements with the second drop.)

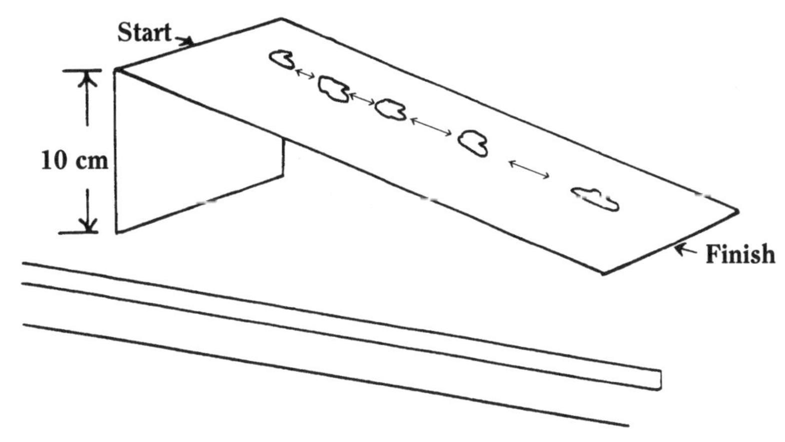

Data table

Distance between drops

Drops	Trial 1	Trial 2	Trial 3
1 & 2 (may be unreliable)	____.__cm	____.__cm	____.__cm
2 & 3	____.__cm	____.__cm	____.__cm
3 & 4	____.__cm	____.__cm	____.__cm
4 & 5	____.__cm	____.__cm	____.__cm
5 & 6	____.__cm	____.__cm	____.__cm
6 & 7	____.__cm	____.__cm	____.__cm
7 & 8	____.__cm	____.__cm	____.__cm
8 & 9	____.__cm	____.__cm	____.__cm
9 &10	____.__cm	____.__cm	____.__cm

9. After you record the measurements for Trial 1, dry off the incline, refill the drip timer, and repeat the procedure. Record your observations for Trial 2 and Trial 3 in the data table.

10. If time permits, raise the incline and repeat the trials. Record your results on a separate sheet of paper.

11. The drip timer (if it is working correctly) releases drops of water at a constant rate. Another way to think about that is to say that a drop of water marks the location of the car about every 1/2 s. *The distance between the drops shows how far the car traveled in about 1/2 s.*

Does the car travel the same distance during each time interval? How can you tell?

____

____

____

GUIDE TO ACTIVITY 12

Determining the Acceleration of a Toy Car

◆What is happening?

This activity is designed to demonstrate another way to find an answer to the general question, "Is the object accelerating?" Watching the car move down the incline would lead most people to answer "Yes" for this case. The car changes from a speed of zero at the top of the ramp to rapid movement at the bottom.

However, in many cases simply watching is unreliable. Using the drip timer or an analogous device allows the observer to use more objective means for deciding whether or not acceleration is occurring. The timer enables us to determine the position of the car at equally spaced time intervals. This allows us to use the following operational definition for acceleration:

> A moving object is accelerating if it travels a *different distance* during each equal time interval that its motion is observed.

For the conditions used in this activity, the distances between the spots become larger for each successive time interval. Therefore, acceleration is taking place.

◆Time management

One class period (40–60 minutes) should be enough time to complete the activity and discuss the results.

◆Preparation

Setting up the mini drip timers is not described in detail in this activity, on the assumption that students will have worked with them before.

Cars used for this activity *must roll freely*; the more easily the car rolls, the better. Use cars with low friction between wheels and axles. Electric cars (such as Stompers) will not work unless they have a "freewheeling" setting that completely disengages the motor from the wheels. Roller skates can also be used for this activity. Simple plastic children's skates with metal axles can be purchased for about $7 per pair in toy stores. Check to be sure that the skates roll smoothly. Laboratory carts manufactured specifically for experiments such as this are ideal, but unfortunately they are expensive and many schools do not have a sufficient number of carts to allow all students to work with them. Hall's carriages are less expensive than larger dynamics carts and are available through many scientific suppliers.

◆Suggestions for further study

A line of drops was produced in the activity that showed the car speeding up. How could a line of drops be produced that would show the car slowing down? Predict how the distance between drops would change as a car is slowing down. Test the prediction.

Have students make a drip timer out of a styrofoam cup and use it to record their own motion as they walk at different speeds on a sidewalk or pavement. Challenge other students to read the drips and determine the direction of travel, points of greatest and least speed, and points of acceleration and constant speed.

◆Answers

2. The car moves slowly at the top of the ramp, and picks up speed as it travels down the ramp.

6. The drops are close together at the top of the ramp. The distance between successive pairs of drops increases as the car moves down the ramp.

8–9. *The observations will vary depending on the car being used and the interval between drops.* The following data are included to show an approximation of what may be observed. The spacings between drops (in cm), starting with drop # 2, were as follows: 3.0, 7.5, 10.0, 13.5, 16.0, 20.5, 24.0.

11. The observations of distance moved between drops are by no means perfect, but they clearly show that between successive pairs of drops the car moves a greater distance.

ACTIVITY 13 WORKSHEET

Using an Accelerometer to Test for Changes in Speed

Materials

Each group will need

- a ruler (or a stick about 30 cm long)
- 15 cm of masking tape
- 40 cm of string
- an index card (3 x 5 or larger)
- a small weight (a 1-ounce lead fishing sinker is ideal)

Vocabulary

- **Accelerometer:** A device used to measure acceleration or determine if acceleration is present (acceleration meter).
- **Operational definition:** A non-mathematical definition of a term that includes an operation to perform and observations to make in order to determine whether the defined phenomenon is present.

◆Background

Making time and distance measurements is not always the most practical way to answer the question, "Is an object accelerating?" This activity provides you with an instrument to determine if acceleration is taking place. Using the instrument involves two steps: you perform an operation and then you make observations. The observations will tell you if acceleration is taking place. We can even define acceleration as the event that is occurring when certain observations are made after a particular operation has been performed. We call this kind of definition an **operational definition.**

◆Objective

To build an **accelerometer** and use it to determine if acceleration is taking place

◆Procedure

1. Draw a center reference line across the card. Tape the card near one end of the ruler, forming a "t" shape. The card's reference line should be centered on the ruler.

Tie the sinker to the string. Turn the ruler so that the t is upside down. Tape the string to the end of the ruler so that the sinker hangs over the card's reference line. (See diagram). This completes your **accelerometer.**

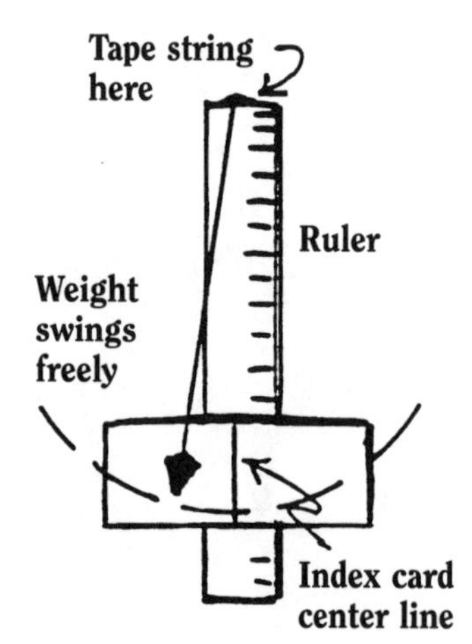

2. Hold the accelerometer out to the side of your body at eye level. The ruler and card should look like an upside down t. Tilt the ruler toward your head slightly so that the sinker can swing back and forth freely without touching the card.

3. Determine the sinker's rest position as follows. Hold the ruler vertically and keep the sinker motionless. If the card and string are properly aligned, the sinker should hang directly over the card's reference line when it is in the rest position. If the sinker does not line up properly, change the position of the string.

4. Be sure that the sinker is not moving. Take a step forward. As you begin to move, observe how the sinker moves away from the rest position. Describe what you observed.

5. Stop the sinker from swinging. Return the sinker to the rest position and take a *very fast* step forward. Compared to the observations you made in step 4, does the sinker move more distance, less distance, or the same amount of distance from the rest position?

6. Hold the sinker in the rest position and begin walking. When you have reached *a constant speed,* release the sinker. Be careful not to push or pull the sinker with your hand as you release it.

When you are walking *at a constant speed,* does the sinker move away from the rest position? Describe your observations below.

7. Write a statement summarizing how the rate of change of speed (acceleration) affects the position of the sinker relative to the rest position.

GUIDE TO ACTIVITY 13

Using an Accelerometer to Test for Changes in Speed

◆What is happening?

The "portable pendulum" assembled for this activity is a simple **accelerometer** (acceleration meter). As the rate of change of speed (acceleration) increases, the deflection from the rest position also increases. At a constant speed, there is *no* deflection from the rest position. The accelerometer only measures *changes* in motion, and indicates the relative *rate of change of speed (the acceleration)* by how far away from the rest position the sinker moves.

Most observers will say that as the person holding the accelerometer changes speeds, the sinker pulls the string to one side or the other. In fact, this is not exactly the case. The sinker's *inertia* (difficult-to-moveness) keeps it motionless at the rest position under two conditions:

• when the accelerometer is not moving, or

• when the person holding the accelerometer is moving at a constant speed in a straight line.

The explanation for why the sinker stays in the rest position for these two cases is given by the first law of motion:

Case 1: An object at rest (*the sinker*) tends to stay at rest (*in the rest position*).

Case 2: An object in uniform linear motion (*the accelerometer's sinker, held by a person walking at a constant speed*) will stay in motion in a straight line and a constant speed (*keeping the sinker at the rest position*) unless acted upon by an unequal force.

When the person holding the accelerometer changes speed (accelerates), the string exerts an unequal force on the sinker. This unequal force causes the sinker to move in the direction of the acceleration. The sinker does not immediately move, however, because of its inertia (difficult-to-moveness). The sinker, therefore, exerts a force against the string that causes the string to move away from its perpendicular rest position.

When the person holding the accelerometer increases speed and moves forward, the sinker stays approximately at the rest position until the string pulls it forward.

When the person slows down, the sinker continues to move forward at the person's previous (higher) speed. It would continue moving at that speed, but the string exerts an unequal force on the sinker to prevent it from flying away from the accelerometer.

There are many possible designs for accelerometers. All of them have one feature in common: a fairly large mass that is free to move relative to a support structure.

Accelerometers provide an operational definition of acceleration: *Acceleration is taking place if the accelerometer's indicator moves away from the rest position.* While this is not a complete explanation of acceleration, it does allow the observer to answer the question "Is acceleration taking place?" Once students can tell whether or not acceleration is occurring, they can begin to relate the scientific definition to everyday examples of this phenomenon.

◆Time management

One class period (40–60 minutes) should be enough time to complete the activity and discuss the results.

◆Preparation

When you introduce the concept of acceleration to your classes, you may wish to give students these three operational definitions of acceleration:

• Acceleration is occurring if an accelerometer's indicator moves away from the rest position.

• Acceleration is occurring if an object travels *different distances* during every equal time interval that its motion is observed.

• Acceleration is occurring if there is a *change in the time interval* required to move a constant distance.

These are not complete definitions of acceleration, but they do provide concrete, non-mathematical methods of determining whether or not acceleration is occurring based on direct observations of moving objects.

◆Suggestions for further study

Take this device on a bus or car and observe what happens when the driver steps on the gas or applies the brakes. What would you *predict* will happen?

◆Answers

4. The sinker moves away from the rest position on acceleration. It moves in the direction *opposite* to the direction that the person holding the accelerometer moves on acceleration, and in the same direction the person moves on deceleration (negative acceleration, slowing down).

5. The sinker moves *more distance* from the rest position when a fast step is taken. The faster the step, the greater will be its displacement from the rest position.

6. At a constant speed, the sinker stays in the rest position (it may move slightly in response to the irregularities of motion in walking).

7. The experimenter will observe *more displacement* of the sinker from the rest position the *greater* the rate of change of speed. In other words, the greater the acceleration, the greater the response of the accelerometer.

ACTIVITY 14 WORKSHEET

Using "Tappers" to Investigate the Fall of a Thread

Materials

Each group will need
- a meter stick
- 4.5 m of sewing thread (preferably heavyweight, white)
- a felt tip pen
- 10 "split shot" fishing sinkers (about the size of a pea)
- pliers
- an aluminum pie pan or cookie sheet

◆Background

Acceleration is occurring if either there is a change in the time interval required to move a constant distance, or if there is a change in the distance traveled during a series of equal time intervals. In this activity we'll devise another method for using these two operational definitions to answer the question, "Is an object accelerating?"

◆Objective

To determine if a falling thread is accelerating

◆Procedure

1. Assemble "tapper threads" consisting of split shot spaced along pieces of thread at the intervals shown in the following diagrams:

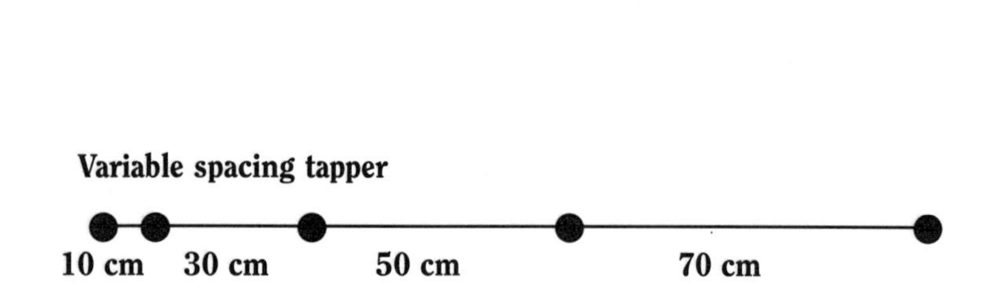

Construct an equal spacing tapper thread as follows:

Cut off a 2.5-m length of thread.

Use the pliers to clamp a split shot at the end of the thread (split shot #1).

Starting at split shot #1, mark off four 50-cm intervals. (Hint: Try not to stretch the thread when you straighten it to measure the 50-cm intervals.)

Clamp a split shot on top of each mark. You will have thread left over beyond the last split shot.

2. Construct a variable spacing tapper thread as follows:

Cut off a 2.0-m length of thread.

Use the pliers to clamp a split shot at the end of the thread (split shot #1).

Starting at the split shot #1, mark off four varying intervals:

Interval 1: 10 cm from split shot #1.

Interval 2: 30 cm from the mark for the second split shot.

Interval 3: 50 cm from the mark for the third split shot.

Interval 4: 70 cm from the mark for the fourth split shot.

Clamp a split shot on top of each mark. You will have thread left over beyond the last split shot.

3. Hold the end of the equal spacing tapper thread so that all the split shot are directly above the pan. Split shot #1 should be about 5 cm above the pan. (If necessary, stand on a chair or table so that you can hold the entire length of string directly above the pan.)

4. Release the thread. *Listen to the intervals between the taps* as the split shot hit the pan. You will probably want to repeat this several times to be certain of the sound pattern.

As the string falls, does the time interval between each pair of taps stay the same, decrease, or increase? ______________________________

5. One operational definition of acceleration can be stated: "Acceleration is occurring if there is a *change in the time interval* required for an object to move a constant distance."

Do your observations of the taps made by the equal spacing tapper thread suggest that acceleration was occurring according to this definition? Explain why or why not.

____

____

____

____

6. Repeat steps 4 and 5 using the variable spacing tapper thread. Shot #1 should be about 5 cm above the pan.

As the string falls, does the time interval between taps stay the same, decrease, or increase? ______________________________

7. Another definition of acceleration is "Acceleration is occurring if an object travels *different distances* during each equal time interval that its motion is observed."

Explain how the observations you made with the variable spacing tapper agree with this definition for acceleration.

____

____

____

____

____

____

GUIDE TO ACTIVITY 14

Using "Tappers" to Investigate the Fall of a Thread

◆What is happening?

The goal of this activity is to demonstrate qualitatively what is meant by the following operational definitions of acceleration:

> Acceleration is occurring if there is a *change in the time interval* required to move a constant distance.
>
> Acceleration is occurring if an object travels *different distances* during a series of equal time intervals.

Neither definition represents a complete explanation of why or how much an object is accelerating. Each definition simply provides a way to address the question, "Is acceleration occurring?"

The time intervals involved in this activity (about 0.14 s for the variable spacing tapper) are too short for accurate direct measurements, even using a stop watch. A multiple-exposure photograph using a strobe light or an electronically-operated timer would be required to obtain accurate interval measurements.

However, the rhythmic patterns of taps are sufficiently distinctive that most people can tell whether the intervals between the taps are equal or unequal. The function of the split shot is to make a noise to enable the observer to tell when a particular point on the thread has reached the ground. From this, one can infer whether or not the falling thread is accelerating.

The thread and weights are acted on by the force of gravity. The force of gravity produces acceleration toward the Earth. The opposing force of air resistance tends to reduce the acceleration toward Earth somewhat, but the effects of air resistance are too small to be detectable by ear.

The algebraic explanation for why the shot spaced at unequal distances along the thread hit the pan at equal time intervals requires using the formula:

$$d = (1/2)at^2$$

This equation relates distance, acceleration, and time for accelerating objects. For this activity, the terms in the equation stand for the following:

> d = total distance the shot moves between the starting point and the surface of the pan (in meters)
>
> a = acceleration due to gravity ($\approx 10\ m/s^2$)
>
> t = time interval the shot required to fall from its starting position to the pan

In order to simplify the calculations using this formula, assume the split shot #1 was just touching the pan when the thread was released. Calculating the total distance fallen by each shot and substituting this distance in the equation produces the following table of values for t:

Data table

Split shot	Total distance fallen	Total time of fall
1	0 m (see assumption)	0
2	0.1 m	0.14 s
3	0.1 m + 0.3 m = 0.4 m	0.28 s
4	0.1 m + 0.3 m + 0.5 m = 0.9 m	0.42 s
5	0.1 m + 0.3 m + 0.5 m + 0.7 m = 1.6m	0.56 s

Sample calculation for shot #2:

$d = 0.1$ m

a = acceleration due to gravity (≈ 10 m/s^2)

t = unknown

Substituting in the equation $d = (1/2)at^2$,

$0.1\text{m} = (1/2)\ (10\ \text{m/s}^2)\ t^2$

Rearranging the expression in terms of t^2, $2d/a = t^2$

$t^2 = 2(0.1\text{m})/(10\text{m}/\text{s}^2)$

$t^2 = 0.02\ \text{s}^2$

Using a calculator to take the square root of this expression gives the value for t:

$\sqrt{t^2} = \sqrt{0.02\ \text{s}^2}$

Therefore, $t = 0.14$ s

All the other values for t are multiples of this time interval.

◆Time management

One class period (40–60 minutes) should be enough time to complete the activity and discuss the results.

◆Preparation

You may wish to prepare the tapper threads before class and perform this activity as a demonstration.

◆Suggestions for further study

If you have access to a reel-to-reel audio tape recorder and microphones, record the sound of the tappers hitting the pan. Make your recording at the fastest tape speed (usually 7 1/2 or 15 inches per second), but play the tape back at the slowest tape speed (usually 1 7/8 inches per second). The sounds will be spread out at the slow playback speed, and differences between the equal spacing tapper and the variable spacing tapper will become even more obvious.

◆Answers

4. The taps sound closer together as the thread falls. This means that the time between taps decreases the further the thread falls.

5. The distance between every two split shot is exactly 50 cm. The thread must fall 50 cm between each pair of taps. Since the time intervals between taps decrease, one can infer that acceleration is occurring.

6. The shot on the variable spacing thread hit the pan at a regular, unchanging rate. The time interval between the taps stays the same.

7. No two split shot on this thread are spaced the same distance apart. Since the taps that they make occur at regularly spaced (equal) time intervals, one can infer that the string is traveling a different distance during every equal time interval. Therefore, acceleration is occurring.

MODULE 4

Interactions of Force, Mass, and Acceleration

◆Introduction

• Why does getting hit by a baseball hurt? A Whiffle Ball™ hitting a batter at the same speed bounces off harmlessly.

• How can a dust particle the size of a grain of sand disable a satellite?

• Why does a truck take longer to reach the speed limit when it is heavily loaded than when it is empty?

Measuring speed, velocity, or acceleration allows scientists to describe how an object is moving. However, these measurements are not sufficient to allow scientists to predict what type of motion will occur when that object is pushed by a force. Predicting motion requires an understanding of the relationship between the object's mass and the force acting on it.

Sir Isaac Newton discovered the nature of this relationship. In his work, *Principia*, Newton described it in what is now called the second law of motion:

Force = mass X acceleration

In this module, segments of the *Eureka!* film series provide a quick overview of how the second law of motion integrates the concepts of speed, force, mass, and acceleration. Other activities introduce an additional measure of motion:

Momentum = mass X velocity

◆Instructional Objectives

After completing the activities and readings for Module 4, you should be able to

- state an algebraic definition of momentum [Reading 10]
- demonstrate the relationship between momentum and mass for objects traveling at the same speed [Reading 10, Activity 19]
- use the algebraic definition of momentum to predict the relative amounts of "bashing power" that can be exerted by a given object as its speed is varied [Reading 10]
- state Newton's second law of motion [Activity 19]
- demonstrate how acceleration changes if the force exerted is kept constant and mass is varied [Reading 11]
- use the second law of motion to explain why a massive object (such as a bowling ball) requires a "harder push" to set it in motion than does a less massive object (such as a volleyball) [Activity 20]

◆Preparation

Study the following readings for Module 4:

Reading 10: Momentum

Reading 11: Force and Motion: Newton's Second Law of Motion

◆Activities

This workshop includes the following activities:

Activity 15: *Eureka!* #3—Speed

Activity 16: *Eureka!* #4—Acceleration part 1

Activity 17: A Marble Race: Does the Mass of the Marbles Affect the Results?

Activity 18: Marble Momentum: Mass versus "Bashing Power"

Activity 19: A Mini Tractor Pull

ACTIVITY 15: VIDEOTAPE

Eureka! #3—Speed

◆Background

The title of this *Eureka!* segment is somewhat misleading. Speeds of the objects shown in the film are simply read off a speedometer calibrated in kilometers per hour. Very little information about how to measure an object's speed is presented. Instead, the program uses common objects to show how varying speed, mass, and amount of force affects what Sir Isaac Newton called the "quantity of motion"—**momentum**.

◆Time management

The running time of the videotape is 5 minutes. At least 15 minutes should be allotted to introduce, run, and discuss the videotape. You may wish to play the videotape at the end of a lesson to reinforce the concepts presented.

◆Comments on the videotape

The term *momentum* is not used in this segment, nor is there a specific statement that momentum is equal to mass times the velocity of a moving object. However, by showing a fast-moving tennis ball of mass 50 g that "packs more wallop" than a slow-moving 20,000-g cannon ball, the meaning of momentum is convincingly demonstrated.

This segment of *Eureka!* illustrates that knowing either the mass alone or the speed alone is not enough information to predict the force that a moving object can exert on another object. Additionally, familiar objects are used to show that in order to go from rest to a given speed, a massive object requires more force to achieve that speed than a less massive object accelerating to the same speed, within the same time period.

These general observations on the relationship between mass and the difficulty of changing an object's speed are used to introduce Newton's second law, F = ma, in *Eureka!* #4.

Concept Summary

- "The tendency things have to keep doing what they are already doing is called **inertia**."*
- "**Mass** is a measure of inertia."*
- "But in order to make a thing change what it is doing, you have to use force, and force varies not only with mass, but also with change of speed."*

Vocabulary

- **Momentum:** The product of an object's mass and velocity; Newton called momentum the "quantity of motion." It can be thought of an an object's "bashing power."

Eureka! Produced by TVOntario © 1981.

ACTIVITY 16: VIDEOTAPE

Eureka! #4—Acceleration part I

◆Background

Viewers of this videotape who have ever ridden a bicycle are familiar with the concepts illustrated here. This videotape will place these ideas in the context of the second law of motion and show how the principles of mechanics explain what happens when we ride a bike.

Concept summary

- "The greater the rate of change of speed, the greater the force required."*
- "Another word for rate of change of speed is acceleration."*
- "The more you accelerate, the more force you need."*
- "But force also varies with mass."*
- "So we say that force equals mass times acceleration."*

◆Time management

The running time of the videotape is 5 minutes. At least 15 minutes should be allotted to introduce, run, and discuss the videotape. You may wish to play the videotape at the end of a lesson to reinforce the concepts presented.

◆Comments on the videotape

The bicyclist shown in this *Eureka!* segment performs a series of tasks that demonstrates several implications of the second law of motion. The biker is first asked to decide which bike would require less force to ride: a light modern bike or a heavy old "high wheeler." The biker knows that force is required to overcome inertia and friction. The older bike is more massive, so it has more inertia and friction, and therefore requires more force to move.

Next, he is reminded of the relationship (shown in Program 3) that force varies with mass and *change of speed*. This leads him to the conclusion that it requires less force to pedal at 25 km/hr than at 50 km/hr. The biker finds that it takes force (pedaling requires a lot of push) to get the bike to top speed.

But does top speed therefore require the most force? To the biker's surprise, the answer is no—maintaining top speed requires *zero force!* When he takes his feet off the pedals while going at top speed, he continues to coast at almost the same speed. Only road friction and air resistance slow him down. Remember the final part of the first law of motion: ...an object in motion tends to stay in motion unless acted upon by an outside force.

If a bike is moving at a *constant* 25 km/hr, its acceleration equals zero. By substituting that value into the equation $F = ma$, the mathematical value of the force at a constant speed is shown to equal zero ($F = m \cdot 0 = 0$). This makes mathematical sense, but probably is not enough to convince most students that an object does not require a force to continue moving at a constant speed. They are, of course, correct about motion on Earth; maintaining motion requires a constant input of force in order to overcome the friction between the moving parts of the bicycle, as well as the friction between the bicycle and the air (**air resistance**). The confusion here is that the *F* in $F = ma$ refers to the *net* force acting on the bicycle (including friction), *not* just the single force of the cyclist pushing on the pedals.

However, modern bicycles are extremely efficient machines that approximate frictionless motion better than most everyday objects. Coasting a long distance at top speed without exerting additional force is an experience that all bicycle riders will remember. This experience demonstrates concretely that force is only exerted while an object is changing speed. In other words, all of the force is exerted while the bike is accelerating.

Vocabulary

- **Air resistance:** A force exerted on a moving object opposite to its direction of motion due to the friction between the object and air. Air resistance is also called *drag* or *air friction*.

Eureka! Produced by TVOntario © 1981.

Viewers are reminded of one additional biking experience before the second law of motion is stated in its entirety: you have to pump harder to make a bike reach top speed in a short period of time than you would if you took longer to reach top speed. This refines the concept of acceleration. Until this segment, acceleration has simply meant that something is "speeding up." The inclusion of a time factor into the concept of acceleration demonstrates that acceleration measures the *rate* of change of speed.

After finishing the demonstrations of acceleration, the second law of motion is stated: Force = mass X acceleration.

ACTIVITY 17 WORKSHEET

A Marble Race: Does the Mass of the Marbles Affect the Results?

Materials

Each group will need

- a meter stick
- 2 sections of N-gauge model railroad track
- 5 marbles of the same size
- 15 cm of masking tape
- several books for supporting tracks

◆Background

Do heavy objects fall faster than light objects? Would a bowling ball dropped simultaneously from the same height as a golf ball hit the ground first? This activity examines these questions.

◆Objective

To discover if an object's mass affects the rate at which it rolls down an incline

◆Procedure

1. Set up two sections of track on a smooth, level surface as shown in the diagram. Check to be sure that the tracks are even with each other, are not bent or curved, and have the same slope.

Place a book at a right angle to the tracks so that a marble rolling down a track will hit the book just after reaching the bottom of the incline. This is the impact point.

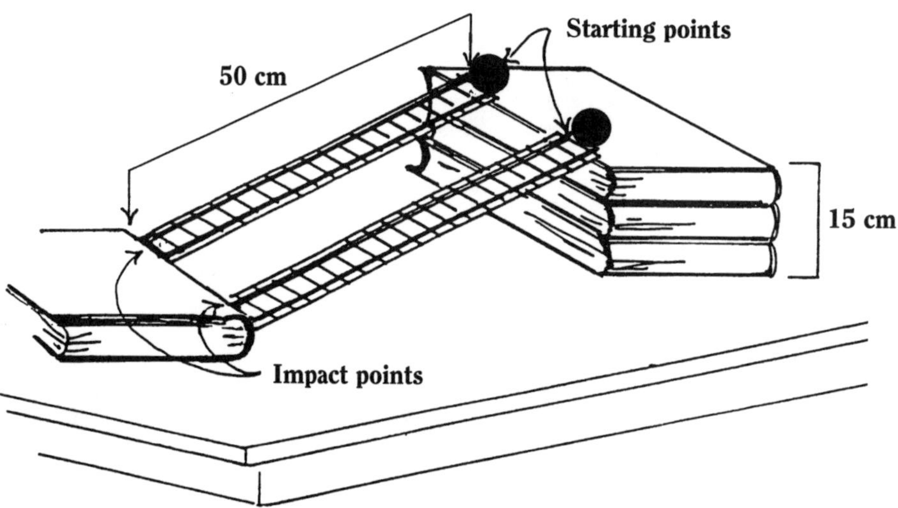

The distance along the track from the starting point to the impact point should be exactly 50 cm. Any height between 10 and 20 cm will produce acceptable results.

Put a small piece of doubled-over masking tape underneath each starting point on the tracks to help hold the tracks in position on the supporting books.

2. Place a marble at the starting point of each track. Release both marbles simultaneously, and listen to them striking the book. (You may want to place a small piece of metal, such as a hinge, in front of the book so that the moment of impact can be heard more clearly.) Repeat this test several times.

Does it sound as though both marbles reach the impact point at about the same time?

____

3. Suppose you released *two* marbles from the starting point of one of the tracks. Would those two marbles roll down the track faster, slower, or at the same rate as a single marble rolling down an identical track?

____

4. Test your prediction: place two marbles at the starting point on one of the tracks, and one marble at the starting point on the other track. Release the marbles simultaneously and listen to them striking the book. Repeat this test several times.

Do two marbles roll down the track faster, slower, or at about the same rate as a single marble?

____

5. Using the procedure described in step 3, test how long it takes *three marbles* to reach the impact point compared to a single marble. Then repeat this procedure using four marbles compared to one marble.

Write a general statement that summarizes the relationship between the number of marbles and the time that it takes for them to roll from the starting point to the impact point.

____

____

____

____

____

6. Each of the marbles has approximately the same mass. Two marbles have about twice as much mass as one marble, and three marbles have about three times as much mass as one marble. How does changing the total mass of marbles change the time required for them to roll 50 cm? Does the time change or not?

____

____

____

____

____

GUIDE TO ACTIVITY 17

A Marble Race: Does the Mass of the Marbles Affect the Results?

◆What is happening?

The goal of this activity is to demonstrate that 1, 2, 3, or 4 marbles released simultaneously from identical starting heights on parallel tracks will reach the *same speed at the bottom of the incline*. The inference that *increasing mass does not change the speed* contradicts what most people assume—that heavier objects fall faster. This incorrect assumption leads to the incorrect prediction that increasing the number of marbles would cause them to roll down faster.

Careful experimenters will observe that marbles released simultaneously hit the book simultaneously. Minor variations in releasing the marbles may cause the marbles to bump each other and travel at slightly different speeds, but in general, the instant of impact of the marbles rolling down the parallel tracks will be practically identical.

This activity does not require the direct calculation of the speed of the marbles in meters per second. Instead, varying numbers of marbles are rolled down parallel tracks simultaneously under identical conditions. The distance traveled from the starting point to the impact point is equal in all cases, and both tracks have the same slope.

The observation that one, two, or more marbles accelerating down the 50-cm length of track always require the same amount of time to reach the bottom leads to this inference: the speed attained by a *single marble* accelerating down the ramp will be the same speed of *two or more marbles* released in the same manner.

Gravitational force causes the marbles to accelerate down the ramp. Acceleration is not measured directly, however, nor is it important to do so in this case; we are only concerned with the inference that the marbles all attain the same speed as they reach the bottoms of the tracks.

Assuming that the surface of the table is level and smooth, no acceleration occurs after the marble reaches the bottom of the track. The first law of motion predicts that the marble would continue moving at a constant velocity after leaving the track if there were no friction or other forces acting upon it. Therefore, it is reasonable to assume that in the short distance between the bottom of the ramp and the impact point on the book, a marble's speed will not change.

◆Time management

One class period (40–60 minutes) should be enough time to complete the activity and discuss the results.

◆Preparation

The results for this activity will vary somewhat depending on how the marbles are released and how accurately they are placed on the track. The sounds of impact should be very close together, however.

The following are some common sources of experimental variation:

- A hump in the track slows the marble, and a dip speeds it up. Mounting the tracks on a smooth board can eliminate this problem.
- A track that moves will cause inconsistent results. Tape down the tracks.
- Accidentally pushing the marbles as they are released gives

inconsistent results. Release the marbles smoothly, without imparting any extra spin.

- Releasing both marbles at exactly the same moment is tricky—try having the experimenter use both index fingers to hold the marbles. This seems to work better for most people than using a thumb and finger on the same hand, or two fingers on the same hand.
- If the book the marbles hit at the bottom of the tracks is not perpendicular with both tracks, the marble on one track will consistently "win the race."

If N-gauge model railroad track cannot be found, the marbles can be rolled down the inside groove of a curtain rod or ruler, as in Activity 18.

◆Suggestions for further study

If you find that students are not persuaded to believe that the acceleration due to gravity is independent of mass, you might try a variation of the activity. Rather than using different numbers of objects with the same mass, roll objects with different masses down an incline. If you have ball bearings about the same size as the marbles, you may want to race them against the marbles. The mass of the ball bearing is much greater than the marble's mass, but both should hit the barrier at about the same time. You may wish to use identical toy cars that can roll freely down an incline. Add mass to one of the cars by taping sinkers to it. It should be clear to students that one car has more mass, yet they will see that both cars still reach the bottom of the ramp at about the same time.

◆Answers

2. Both marbles hit the book at the same time.

3. Answers will vary.

4. Two marbles released simultaneously roll down the track at about the same rate as one marble rolling down the parallel track.

5. The number of marbles does not affect the time required to roll down the track.

6. Increasing the mass does not change the time required to roll 50 cm.

ACTIVITY 18 WORKSHEET

Marble Momentum—Mass versus "Bashing Power"

◆Background

The amount of **momentum**, or "bashing power," that an object has depends on both its mass and its velocity. In this activity we'll see how an object's mass affects its bashing power.

◆Objective

To determine if increasing mass increases bashing power

Materials

Each group will need
- a meter stick
- a grooved ruler
- 3 marbles of the same size
- 15 cm of masking tape
- several books for supporting the ruler
- 1 "slider car" (4 X 12 cm piece of cardboard or stiff paper folded into a V shape)
- a manila folder as a sliding surface (optional)

◆Procedure

1. Set up the ruler on a smooth, level surface as shown in the diagram. Put a small piece of doubled-over masking tape under the 30-cm end of the ruler to help hold it in position on the supporting books.

The distance along the ruler from the starting point to the impact point should be about 15 cm. Supporting books at any height between 10 and 20 cm will produce acceptable results.

2. Place the 0-cm end of the ruler at the edge of the folder. Mark where the point of your slider car rests on the manilla folder with a starting dot. Place a marble at the 15-cm mark (the starting point) and release it.

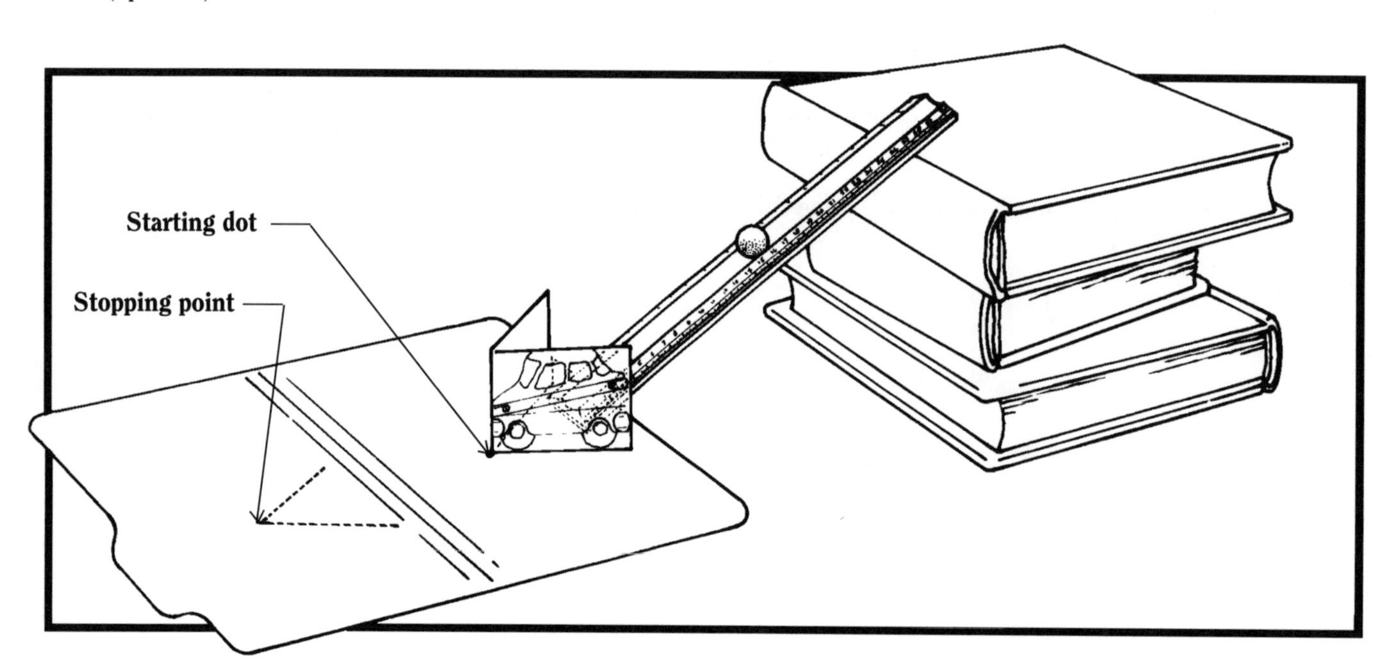

3. Mark the stopping point of the slider with a pencil. Determine how far the car moved by measuring (to the nearest 0.1 cm) the distance between the starting dot and the stopping point. Record your observation in the data table that follows.

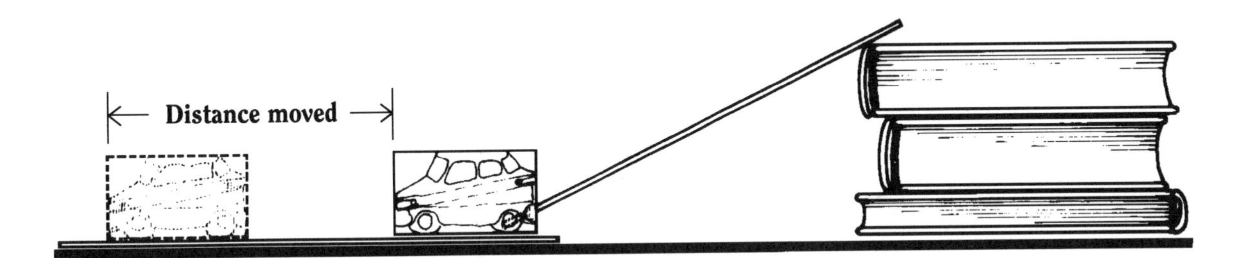

Data table

Distance between starting dot and stopping point

Number of marbles	Trial 1	Trial 2	Trial 3
1	___.__ cm	___.__ cm	___.__ cm
2	___.__ cm	___.__ cm	___.__ cm
3	___.__ cm	___.__ cm	___.__ cm

4. Measure the distance the car moves when it is hit by 2 and 3 marbles. Center the marble groups over the 15-cm mark before release. Make three trial runs with each marble group. Record these measurements in the data table above.

5. Write a statement relating the distance the car moves to the number of marbles hitting it.

____

____

____

____

____

6. Do these results suggest an explanation for why a freight train hitting a car at 30 km/hr does much more damage to a car than would occur if a car were hit by another car at 30 km/hr?

____

____

____

____

____

GUIDE TO ACTIVITY 18

Marble Momentum—Mass versus "Bashing Power"

◆What is happening?

The distance that the car slider moves in this activity depends on how many marbles roll down the ruler and hit it. The distance that the car moves shows the marbles' "bashing power," or **momentum.** The greater the momentum, the greater the distance the car moves.

Activity 17, "A Marble Race," uses a similar apparatus to demonstrate that one, two, three, or more identical marbles always have the same speed at the bottom of the track. It is reasonable to assume that the speed of the marbles hitting the car in this activity will also be the same for each set of trials.

Since the speed of the marbles is constant when one or more marbles hit the car, the observations for the activity lead to the following inference:

For objects moving *at the same velocity*, the object having the *most mass* will also have the *most momentum*, or bashing power.

Another way of stating this relationship is:

For objects traveling at the same velocity, momentum is directly proportional to mass.

For this activity, mass is varied by changing the number of marbles rolling down the ramp. Doubling the number of marbles doubles the mass. Doubling the mass doubles the momentum.

This activity demonstrates the effects of momentum for a special laboratory situation in which each object has the same velocity and mass. The mass is varied from trial to trial by changing the number of objects colliding. However, scientists also study the momentum of objects in the real world with masses ranging from those smaller than atoms to those larger than the Sun, moving at velocities ranging from zero to the speed of light.

Sir Isaac Newton conceived of momentum as the "quantity of motion." He discovered a general mass and velocity relationship for calculating the momentum of any object:

momentum = mass X velocity

which is often written using the symbols

$p = mv$

All that is required to calculate momentum is knowledge of the object's mass and its velocity. Newton used his equation to describe the motion of many objects, ranging from falling apples to orbiting planets.

◆Time management

One class period (40–60 minutes) should be enough time to complete the activity and discuss the results.

◆Preparation

The simple apparatus used for this activity produces somewhat variable data. However, using good technique, results that are consistent enough to be convincing will be obtained. Most experimenters will conclude that increasing the number of marbles increases the bashing power. Analyzing sources of experimental variation is a valuable experience for students.

N-gauge model railroad track may be used in place of the grooved ruler, similar to Activity 17. But if the railroad tracks span more than 50 cm between the starting point and the impact point, they tend to bow. Upward or downward bends in the tracks affect the results. Measure and adjust the starting point so that the distance is correct, and be sure that the track is straight. One optional way to ensure that the tracks remain in good position is to tape each track to a board or meter stick before performing the activity.

There are several reasons for the variations in distance that the car moves. More force is "used up" initially to overcome the frictional forces that hold the slider in place than is used to maintain the slider's motion against friction once it has started sliding and eventually slides to a stop.

Other factors producing non-uniform motion include bumping and

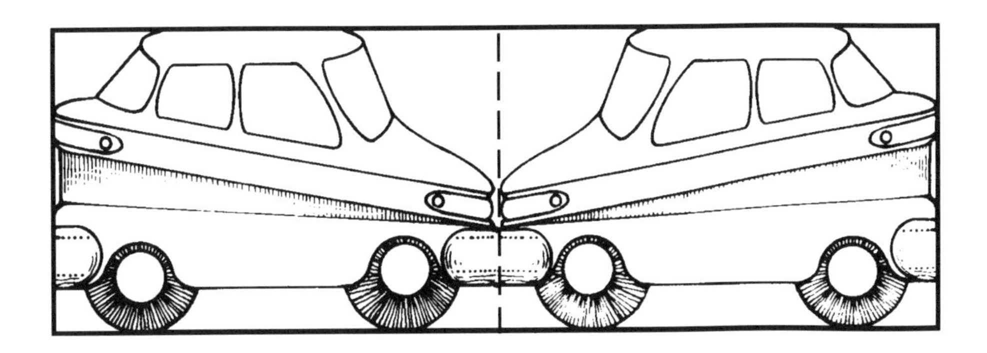

Slider car

rubbing within the groups of marbles after they are released. This causes them to spread out along the ruler, and may cause the marbles to hit the car slider unevenly.

Variation is also caused by the way the marbles are released. The last marble in a line of three starts higher on the track than the first, and travels an extra 2 cm or so down the ruler. The additional speed gained from rolling the extra distance is partially canceled out by collisions with other marbles on the track and by the shorter distance rolled by the first marble.

This recitation of possible sources of error for the activity may sound discouraging, but take heart: results obtained using ultra-low-friction collision devices costing hundreds of dollars vary also. This activity will help your students see the big picture of the concept of momentum. Do not worry about getting results with "three significant digit" accuracy.

◆Suggestions for further study

How does ramp height (which causes the marbles' speed to vary) affect the bashing power of the marbles? After collecting data on the marbles for a given height of the ramp, lower the ramp and repeat the experiment. Raise the ramp to a level above the original height setting, and repeat the experiment.

How does mass affect the bashing power of a ball? Repeat the experiment using a ball bearing the same size as the marbles. If you can obtain a small ball made of plastic or some other lightweight material, test its bashing power also.

Determine the sliding distance for two marbles released at 15 cm. Where do you have to put *one* marble on the ruler to get the same sliding distance? Where do you have to put a three marble group to get the same sliding distance?

◆Answers

4. Answers will vary, depending on the slope of the track and the type of car used.

5. The more marbles hitting the car, the greater the distance that the car moves.

6. The freight train possesses much more mass than a car. Therefore, at comparable speeds, the train has much more bashing power than a car, and will do much more damage to any object that it strikes.

ACTIVITY 19 WORKSHEET

A Mini Tractor Pull

◆Background

How does the mass of an object affect its behavior when a force acts on it? This is another question Newton's second law of motion can help answer.

◆Objective

To learn how increasing an object's mass affects how much a force can cause the object to accelerate

◆Procedure

1. Use masking tape to mark start and finish lines 1 m apart on a table top or smooth floor.

2. Place the plastic cup on the cargo area of the tractor. Use masking tape to attach the cup securely in place.
Tie or tape a piece of string between the tractor and the trailer to serve as a towline.

3. Place the front bumper of the tractor even with the starting line. Turn on the motor and release the tractor.
Determine how many seconds it takes the tractor to reach the finish line.

Materials

Each group will need

- a meter stick
- a battery-operated electric model car or truck (the "tractor")
- a model railroad car (the "trailer")
- 20 cm of string
- cargo—60 pyramid-type fishing sinkers, 2-ounce size (or 30 laboratory masses, each having a mass of 100 g)
- a large (475-mL) plastic cup
- a timer or clock with a second hand
- masking tape

Hint: *Be sure the trailer is directly behind the tractor each time before releasing it or it may veer to one side and cause inconsistent results.*

Hint: *If the tractor is unable to pull 3 kg of cargo to the finish line, add smaller amounts of cargo to the cup for each trial. For example, you might decide to add cargo in increments of 0.25 kg (5 lead sinkers) per trial. If you must change the procedure, record the actual amount of mass that you add for each trial in the data table.*

Data table

Seconds required for tractor to move 1 meter

	Mass of cargo added to trailer (kilograms)						
	0	0.5	1.0	1.5	2.0	2.5	3.0
Trial 1	___s	___s	___s	___s	___s	___s	___s
Trial 2	___s	___s	___s	___s	___s	___s	___s
Trial 3	___s	___s	___s	___s	___s	___s	___s
Average time	___s	___s	___s	___s	___s	___s	___s

4. Repeat the towing trial two more times using the same procedure. Record your observations in the data table.

5. Load 0.5 kg (approximately 20 ounces) of cargo into the cup attached to the trailer. Ten lead fishing sinkers each weighing 2 ounces have a total mass of approximately 0.5 kg.

Place the tractor even with the starting line and measure the time it requires to reach the finish line. Record your observations in the data table.

Fill in the data table by repeating the towing trials and adding the amounts of cargo called for.

6. For each quantity of cargo, calculate the average time it takes for the tractor to move 1 m. Record these averages at the bottom of the data table.

7. What supplies the force (pull) to move the trailer and the cargo?

____

____

____

____

8. Write a statement that describes how adding mass to the trailer changes the time required for the tractor to tow its cargo 1 m.

____

____

____

____

____

9. Have you ever seen a large bus or truck accelerating from a dead stop up to highway speed? Use your observations for this activity to explain why trucks stopped on the road take so long to reach the speed limit.

____

____

____

____

____

GUIDE TO ACTIVITY 19

A Mini Tractor Pull

◆What is happening?

This activity uses a model tractor-trailer to relate the second law of motion to everyday experiences with accelerating objects. Most experimenters will conclude on the basis of their observations that *as the mass of the trailer increases, its acceleration decreases*. This is the result predicted by F = ma, the second law of motion.

Experimenters are not required to determine precise numerical values for force, mass, or acceleration.

Mass is added to the trailer in equal increments for each successive trial. Determining the total mass (in grams) of the tractor, trailer, and load is not essential.

Acceleration is not measured directly. The time it takes the tractor to move one meter is used to estimate its acceleration. The longer the time interval, the smaller the acceleration; conversely, shorter time intervals for moving one meter show a greater acceleration.

Force for moving the trailer and load is produced by the motor of the tractor. The motor always operates at its maximum output; it cannot produce additional force. Therefore, the total amount of force applied to the trailer in this activity remains constant for all the trials. Since the force remains constant for this activity, the tractor-trailer's motion can be analyzed using a simplified statement of the second law of motion. When a constant force is acting, the equation Force = mass X acceleration can be restated:

constant = mass X acceleration

Rearranging the terms in the simplified equation for the second law of motion shows the following relationship between acceleration and mass:

$$\text{acceleration} = \frac{\text{constant}}{\text{mass}}$$

In other words, when a constant force is acting, *acceleration is inversely proportional to mass*: As mass increases, acceleration must decrease.

This simplified statement of the second law of motion predicts that a lightly-loaded vehicle will require less time to accelerate to its top speed than a similar vehicle that is heavily loaded.

For this activity, most experimenters find that a heavily-loaded tractor-trailer requires more time to travel one meter than an empty tractor-trailer. A longer time interval implies that less acceleration is taking place with the heavy load. The observations of the tractor-trailer agree with the results predicted by the second law of motion.

◆Time management

One class period (40–60 minutes) should be enough time to complete the activity and discuss the results.

◆Preparation

Before having students perform this activity, test the towing ability of the electric cars that you plan to use as tractors. Many different types of model cars are manufactured; some lack the force to tow the amount of weight called for in the activity. Stomper™ model cars work very well as tractors for this experiment. If the electric cars that you use have more than one speed setting, be sure to use the same speed setting for all trials.

If the car cannot tow a cargo of 3 kg, have students use smaller increments of mass when loading the trailer. If possible, select masses that are small enough to allow students to add mass to the trailer at least three times without causing the tractor to stop moving.

If the car spins its wheels while towing the trailer, tape a 1- or 2-ounce lead fishing sinker on top of it. The sinker should be in place for all runs, including the zero cargo run.

An easy way to attach the tow line to the tractor and trailer is to stick the string in place using masking tape or duct tape.

Depending on the heights and shapes of the tractor and trailer, you may wish to make this a "tractor *push*" rather than having the tractor pull the cart. The pushing technique gives similar results, and eliminates the necessity of using a tow line.

Anything that rolls smoothly in a straight line and is capable of carrying cargo may be used as the trailer. Inexpensive plastic children's roller skates or toy trucks are possible substitutions. Laboratory testing carts such as Hall's carriages are ideal. They are available through many scientific suppliers, but their cost may prove prohibitive.

Almost any set of objects that have similar masses can be used as cargo, as long as these objects will fit on the trailer. Cans of soup or other canned items make good cargo. Full aluminum soft drink cans each have a mass of about 350 g. The trailer that you use must be large enough to accommodate 3 or more cans if you wish to use them in place of lead sinkers.

Gravel, if it is relatively uniform in size, can be used as an inexpensive (or free) alternative to sinkers or laboratory masses.

Because of the amount of cargo needed for each group, you may wish to perform the activity as a demonstration.

◆Suggestions for further study

Stage a tractor pull contest using electric toys provided by students. Which toy can pull the most mass? Does the number and size of batteries affect the results? In the activity, the same force is applied to different masses. Try using the same materials to apply different forces to the same mass. Try having two tractors pull a trailer and compare the resulting acceleration to that produced by one tractor pulling the trailer. When using two tractors, should they be arranged in a train-like or side-by-side fashion?

◆Answers

3–6. The exact results will vary. The following data were obtained using a Stomper as the tractor, a Hall's carriage as the trailer, and adding 0.5 kg of lead sinkers to a cup on the trailer for each trial:

Average times that tractor, trailer, and cargo required to move 1 m:

0 g of cargo: 3.0 s

500 g of cargo: 3.8 s

1000 g of cargo: 5.2 s

1500 g of cargo: 6.5 s

2000 g of cargo: tractor could not move

Using a different Stomper as the tractor, a homebuilt lab cart for the trailer, and placing 10-ounce cans of soup (each can having a mass of about 300 g) on the lab cart gave the following times:

0 cans as cargo: 3.0 s

1 to 6 cans of soup as cargo: 3.8 s, 4.7 s, 5.7 s, 7.3 s, 8.1 s, 10.4 s

7. The motor of the tractor, powered by an electric battery, supplies the force to move the trailer.

8. As the mass on the trailer increases, the time required to tow the trailer one meter also increases.

9. For a given amount of force (such as a truck engine produces), a larger mass requires more time to accelerate to the same speed than does a smaller mass. Emptying a truck does not affect the force that its engine can produce. However, since the engine of an empty truck is exerting its force on a smaller mass, the truck will have a greater acceleration when it is empty.

MODULE 5

Applying the Laws of Motion

◆Introduction

• Why does a shotgun kick the shoulder of the person firing it?

• How can rockets accelerate in the vacuum of space, where the rocket exhaust has no air to push against?

• What prevents a roller coaster doing a loop from falling off the tracks?

This module will introduce the third law of motion and extend the application of the second law of motion to situations that are more complex than those encountered in previous modules. Many more types of motion can be explained when Newton's third law is understood. Newton's third law of motion may be stated either in terms of forces or in terms of actions and reactions:

For every force there is an equal and opposite force.
To every action there is an equal and opposite reaction.

This module will demonstrate that the second and third laws when used together form a powerful way to investigate motion.

◆Instructional Objectives

After completing the activities and readings for Module 5, you should be able to

- state Newton's third law of motion in two ways:
  1. in terms of action and reaction pairs, and
  2. in terms of opposing forces [Reading 12]
- explain why a gun kicks when a bullet is fired [Activity 20]
- see that the force of gravity acting on an object that is resting on a surface is balanced by the force that the structure exerts on the object [Activity 21]
- see that acceleration occurs if the opposing forces acting on an object are not balanced [Activities 21 and 22]

◆Preparation

Study the following reading for Module 5:

Reading 12: The Third Law of Motion

◆Activities

This module includes the following activities:

Activity 20: Kodak Cannons (or, Investigating the Motions of Action-Reaction Pairs)

Activity 21: Fettucini Physics (or, Studying Equal and Opposite Forces in Support Structures)

Activity 22: Hair Drier versus Gravity—Equal and Opposite Forces?

ACTIVITY 20 WORKSHEET

Kodak Cannons (or Investigating the Motions of Action-Reaction Pairs)

The container *must* be test fired prior to the class demonstration. Test firing instructions are found in the "Preparation" section of the "Guide to Activity 20."

◆Background

Newton's third law states that for every action there is an equal and opposite reaction. In this activity, your teacher will demonstrate this with a rather messy example.

◆Objective

To demonstrate Newton's third law

Materials

Each group will need

- safety glasses or goggles for each participant and observer
- a meter stick
- 3 empty Kodak™ 35-mm film containers
- 4 Alka-Seltzer™ tablets
- 2 lead weights (1- or 2-ounce fishing sinkers)
- 70-cm section of N-gauge model railroad track
- 100 ml of room-temperature water
- a sponge or towel

◆Procedure

This activity is a demonstration! Steps 1–5 are to be performed by the teacher while all students remain at least 3 m away from the track. Remember to test fire the container prior to the demonstration. Please refer to the "Guide to Activity 20" before performing this activity.

1. Place the section of track on the floor or demonstration table. Be sure that the students are at least 3 m from the track. Neither end of the track should point at any person or any breakable object.

2. Put the cap on an empty film container, and place the container on the track so that it can slide freely along the track. The cap and the bottom of the container should fit neatly between the rails. This will be the bullet. The other container will be the cannon.

The firing positions for the bullet and the cannon are shown in the diagram below:

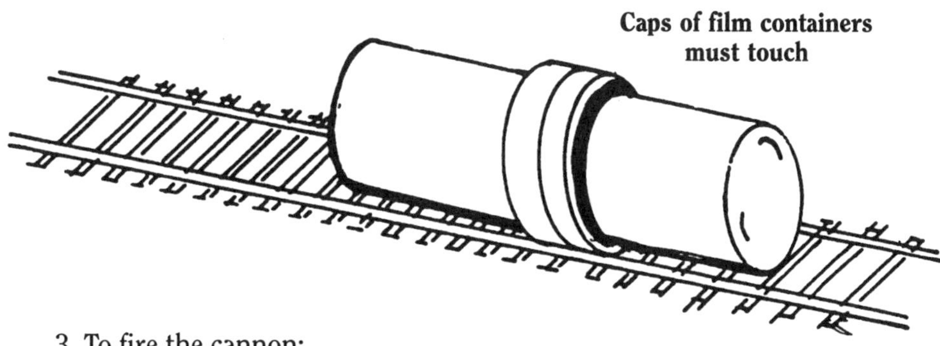

No students should be involved in the firing procedure (steps 1–5). This activity is to be demonstrated by the teacher.

Everyone observing this activity should wear safety goggles and keep at least 3 meters away from the track.

3. To fire the cannon:

Place the empty bullet on the center of the track.

Load the cannon with Alka-Seltzer and water as you did for test firing. *Quickly* place it on the track so that the top of the cannon touches the top of the bullet as shown in the diagram. *Step back, and be sure that no one is near the cannon.*

After the cannon fires, measure the distances that the bullet and the (bottom of the) cannon moved. Record your results in the data table.

If the cannon does not fire within 2 minutes, *carefully* pick it up. *Hold the top and bottom firmly, and point the top away from everyone.* Pull open the top and release any remaining pressure. Using a different film container, repeat the firing.

4. Completely fill the bullet with water, seal the top, and place it back on the center of the track. Observing all safety rules, fire the cannon as you did in step 3, and measure the distances that the bullet and the cannon bottom travel. Record your results in the data table.

Data table

	Empty Bullet	Bullet + Water	Bullet + Lead
Distance cannon traveled	____m	____m	____m
Distance bullet traveled	____m	____m	____m

5. Place a lead weight inside the bullet and reseal the top. Observing all safety rules, repeat the measurements done for steps 3 and 4.
Clean and dry the floor!

6. What provides the force that moves the cannon and bullet?

____

____

7. When the bullet is empty, it has about the same mass as the cannon. Do the cannon and bullet travel about the same distance from the starting position in this case? In what directions do they move?

____

____

8. Explain how your observation of the cannon and empty bullet support the statement, "For every action there is an equal and opposite reaction."

____

____

____

9. The water-filled and lead-filled bullets were more massive than the cannon. Describe the motion of these two types of bullets compared to the motion of the cannon.

____

____

____

10. Real cannons are always much more massive than the projectiles that they fire. Use your observations to explain why a cannon must have more mass than the projectile that it fires.

____

____

____

GUIDE TO ACTIVITY 20

Kodak Cannons (or Investigating the Motions of Action-Reaction Pairs)

◆What is happening?

This activity is designed to demonstrate Newton's third law of motion: *For every action there is an equal and opposite reaction.*

The cannon and bullet (the film containers) are (for all practical purposes) identical. The empty bullet is slightly less massive than the loaded cannon, but when the top comes off of the cannon, the small difference in mass is essentially canceled out.

The force (provided by the carbon dioxide produced by the Alka-Seltzer) is transmitted equally to both the cannon and bullet. Since the *same force* is acting on both the bullet and the cannon, and their *masses are equal*, we can use their motions to illustrate the third law of motion.

The bullet and cannon both move about the same distance; therefore we can infer that the action and reaction forces have the same magnitude. The bullet and cannon move in opposite directions; therefore we can infer that the action and reaction forces act in opposite directions.

The track serves to keep the film containers from rolling, and makes the direction of the motion more obvious. However, similar results can be obtained by placing the film containers directly on a table or smooth floor.

Changing the mass of the bullet demonstrates another feature of action-reaction pairs: if one of the objects acted on by a force is much more massive than the other, the distances that the two objects move will be different. The more massive object will not move as much as the less massive object of the pair.

◆Time management

One class period (40–60 minutes) should be enough time to complete the activity and discuss the results.

◆Preparation

Some film containers (maybe 1 out of 4) will not seal tightly enough to function as cannons. The only way to tell whether or not a particular container will work is by test firing it. Test firing the cannon before your students arrive helps insure that the demonstration will proceed smoothly.

Test fire the cannon as follows:

1. *Put on your safety goggles!* Accidents can happen to teachers, too. Good scientists wear eye protection!

2. Pour enough room-temperature water in the film container to cover the bottom to a depth of about 0.5 cm.

3. Place a small piece of Alka-Seltzer (about 1/4 tablet) into the film container.

4. *Quickly* place the cap on the film container. *Hold the cap firmly in place*. Shake the container briefly (3 seconds or less), then place it *cap down* on a table.

5. *Quickly* step back 1 meter from the table. *Be sure that no one gets near the cannon until it fires!*

If the film container does not pop apart within 2 minutes, *carefully* place your hand on top of it, and *while holding both the top and the bottom firmly,* pull the edge of the top up a small amount to release any remaining pressure. Repeat the test-firing procedure using different film containers until you find one that will shoot the bottom of the container at least 1 m high.

The probability of being injured while performing this activity is extremely low, but treating the apparatus as though it were a real loaded cannon is strongly recommended. Always use some form of eye protection during this activity.

The Kodak brand film containers used for this activity are light, flexible, and almost impossible to shatter. The velocities that they achieve are fairly low. However, it is strongly recommended that no fuel other than Alka-Seltzer be used for the following reasons:

1. The tablet must partially dissolve in order to release a significant amount of carbon dioxide gas. The time required to dissolve the tablet allows the experimenter to back away from the apparatus.

2. If an observer removes his safety goggles and somehow gets splashed in the eye by the Alka-Seltzer, no chemical burn is likely. However, rinsing the eye with water is recommended. Consult a doctor in case of injury.

Camera stores and film processing labs are usually very willing to give you (without charge) all of the film containers that you will need for classroom use. A busy film processing operation throws hundreds of them into the trash every day.

Brands other than Kodak may possibly work, but are not recommended. The Kodak containers are the sturdiest, best sealing containers available.

◆Suggestions for further study

One can also use the second law of motion to predict the motions of the bullet and the cannon.

For this activity, the *same force* acts on both the bullet and the cannon. Using the second law of motion, F = ma, the force acting on the bullet and the cannon can be stated as follows:

$$\text{Force} = \text{mass}_{\text{bullet}} \times \text{acceleration}_{\text{bullet}}$$

and

$$\text{Force} = \text{mass}_{\text{cannon}} \times \text{acceleration}_{\text{cannon}}$$

Since the same force (the explosive release of the gas pressure produced by the Alka-Seltzer) acts on both the bullet and the cannon, the products of mass and acceleration in both equations are equal. Therefore,

$$\text{mass}_{\text{bullet}} \times \text{acceleration}_{\text{bullet}} = \text{mass}_{\text{cannon}} \times \text{acceleration}_{\text{cannon}}$$

This equation shows that when the mass of the bullet and the cannon are *equal,* they will have the same acceleration. If the masses are *different,* the more massive member of the pair (in this activity, the bullet containing water or lead) will have a smaller acceleration.

Although acceleration is not measured directly for this activity, the distances moved are much smaller for the more massive bullets than for the cannon; this correlates with the predicted pattern of motion.

◆Answers

3. The exact distances that the film containers move vary, but the *general pattern* is as follows:

• The empty bullet travels about the same distance as the cannon.

• The bullet containing water travels a shorter distance than it did when empty, but the cannon acting against it travels a greater distance than before.

• The bullet containing the sinker will hardly move at all; the cannon acting against it will move a longer distance than it does for the two previous sets of conditions.

6. The gas (carbon dioxide) released as the Alka-Seltzer dissolves in water provides the force.

7. The cannon and the bullet move about the *same* distance, but in *opposite* directions.

8. Both the cannon and the bullet have about the same mass. They move equal distances but in opposite directions. The fact that the action and reaction forces are equal is shown by the *equal distances* that the film containers move; the fact that the action and reaction are opposite is shown by the *direction* in which they move.

9. The more massive bullets moved less distance than the cannon. As the mass of the bullets increased, the distance that the cannon moved increased.

10. If a real cannon had the same mass as its projectile, the cannon would move backward (toward the people shooting it) as fast as the projectile moved forward. There is not much military advantage to shooting a cannonball at someone else if the cannon is going to fly back and hit you as hard as the cannonball hits them.

ACTIVITY 21 WORKSHEET

Fettucini Physics (or Studying Equal and Opposite Forces in Support Structures)

◆Background

An object is supported by a structure when the downward force of gravity on the object is balanced by the upward force exerted by the structure. In this activity, you'll attempt to build a structure that will provide a force equal and opposite to the force exerted by gravity on a stack of books.

Materials

Each group will need

- a meter stick
- 20 pieces of uncooked fettucini (spaghetti may be used, but all groups should use the same type of pasta)
- 1 m of masking tape
- a heavy book (all groups should use books weighing the same)

◆Objective

To demonstrate that the downward force of gravity on a stack of books has to be balanced by the force exerted on the books by the structure on which they rest

◆Procedure

1. Your task for this activity is to design and build a structure or group of structures that will exert an *upward force* on the books equal to the *downward force* exerted on the books by the Earth. Only the uncooked fettucini and masking tape may be used in assembling your structures.

2. Design requirements:

The completed structure must support the book a *minimum of 5 cm* above the table.

You are not required to use all the pasta and masking tape, but no additional tape or pasta may be used.

! Wear safety goggles or other eye protection when placing books on your structure. The pasta could produce sharp fragments when shattered.

3. Before placing a book on the completed structure, sketch your design in the space below.

4. Determine the maximum amount of upward force that your structure can exert by placing additional books on top of it one at a time.

How many books can your structure support?

___

5. What happens when the *upward force* from the pasta structure is *smaller* than the *downward gravitational force* on the stack of books?

___

___

___

___

6. Carefully remove the stack of books from your broken pasta structure. Can you tell which parts of the structure broke first? Does this suggest a way to make a similar structure stronger?

___

___

___

___

___

___

___

___

7. Please clean up the pieces of pasta and dispose of them properly.

GUIDE TO ACTIVITY 21

Fettuccini Physics (or Studying Equal and Opposite Forces in Support Structures)

◆What is happening?

This activity shows what happens when the *downward* gravitational force on an object exceeds the *upward* force on the object exerted by the structure on which the object rests. As long as the opposing forces are balanced there is no net force acting on the object and the *F* in F=ma is zero. Since the mass is not zero, in order for both sides of the equation to be the same the acceleration must be zero. So, when the forces are equal and opposite there is no change in motion.

! If you have students perform a task similar to this using materials other than pasta, place limits on the amount of material that can be used. Some future engineers may possibly build a structure so strong that testing the structure becomes dangerous because of the heavy weight it can support without breaking. Don't let anyone get their fingers underneath a structure that is about to collapse.

Let's examine the forces acting on the book. There is an *upward* force that the fettucini structure exerts, and the *downward* force of the Earth acting on the book (we call this downward force the *weight* of the book).

Since the book is at rest, its acceleration is zero. According to the second law of motion, the *net force* acting on the book must also be *zero*. Therefore, the upward force exerted by the fettucini structure supporting the book is *equal in magnitude* and *opposite in direction* to the weight of the book. These two forces acting on the book are not the action and reaction forces of the third law since they act on the same object.

The two forces involved that are described by the third law are the force the book exerts on the structure and the force the structure exerts on the book. The *upward* force exerted on the book by the fettucini structure is equal in magnitude and opposite in direction to the *downward* force exerted by the book on the structure. In terms of magnitudes, the force that the fettucini structure exerts on the book is equal to the force the book exerts on the fettucini structure.

Combining these statements based on the second and third laws of motion, we can conclude that the force of the book on the fettucini structure equals the weight of the book.

The book falls when the upward force exerted by the pasta structure is less than the weight of the book (gravitational force acting on the book). The excess downward force breaks the structure, and the book accelerates toward the center of the Earth until it strikes some other surface (such as a table or the floor) that can exert enough upward force to balance the downward force of the book. When the book is at rest, the opposing forces on the book must be equal.

◆Time management

One class period (40–60 minutes) should be enough time to complete the activity and discuss the results. You may wish to have the students prepare their structures (or just their designs) out of class so that they will be encouraged to be creative and to take the time to experiment.

◆Preparation

In this activity, students may learn more from structures that *do not* perform as expected than from structures that satisfy the criterion of holding up a book.

Allow the students time to experiment on their own with the materials. Encourage them to find their own solutions using radical designs and creative thinking, rather than building some standard structure to a particular set of specifications.

◆Suggestions for further study

You may wish to assign this activity as a small project for students to work on over a period of time. You may even want to assemble several classes to have a competition among their designs. Give prizes for the strongest, best-looking, and most unusual support structures entered in the contest.

There are endless variations to the general task of building a support structure. Materials you might consider using include matchsticks, egg cartons, soda straws, cardboard, newspaper, etc. Pasta was chosen in this case because it is cheap and readily available.

◆Answers

3. There is no single correct answer for how to design a structure out of tape and uncooked pasta to support a maximum number of books. One design that works is shown below. There are many possible designs that will work. *Do not* use this design only.

4. The number of books supported varies. The design shown held up three medium-sized books.

5. When the upward force of the structure is smaller than the downward force on the books, the books fall and break the structure.

6. Answers will vary. The pasta resists compression very well, but breaks very easily when it is twisted or bent. Strong structures transfer the weight of the books to the table without causing the pasta to bend. Tape wrapped around the fettucini helps to prevent bending.

A solution for the fettucini physics problem

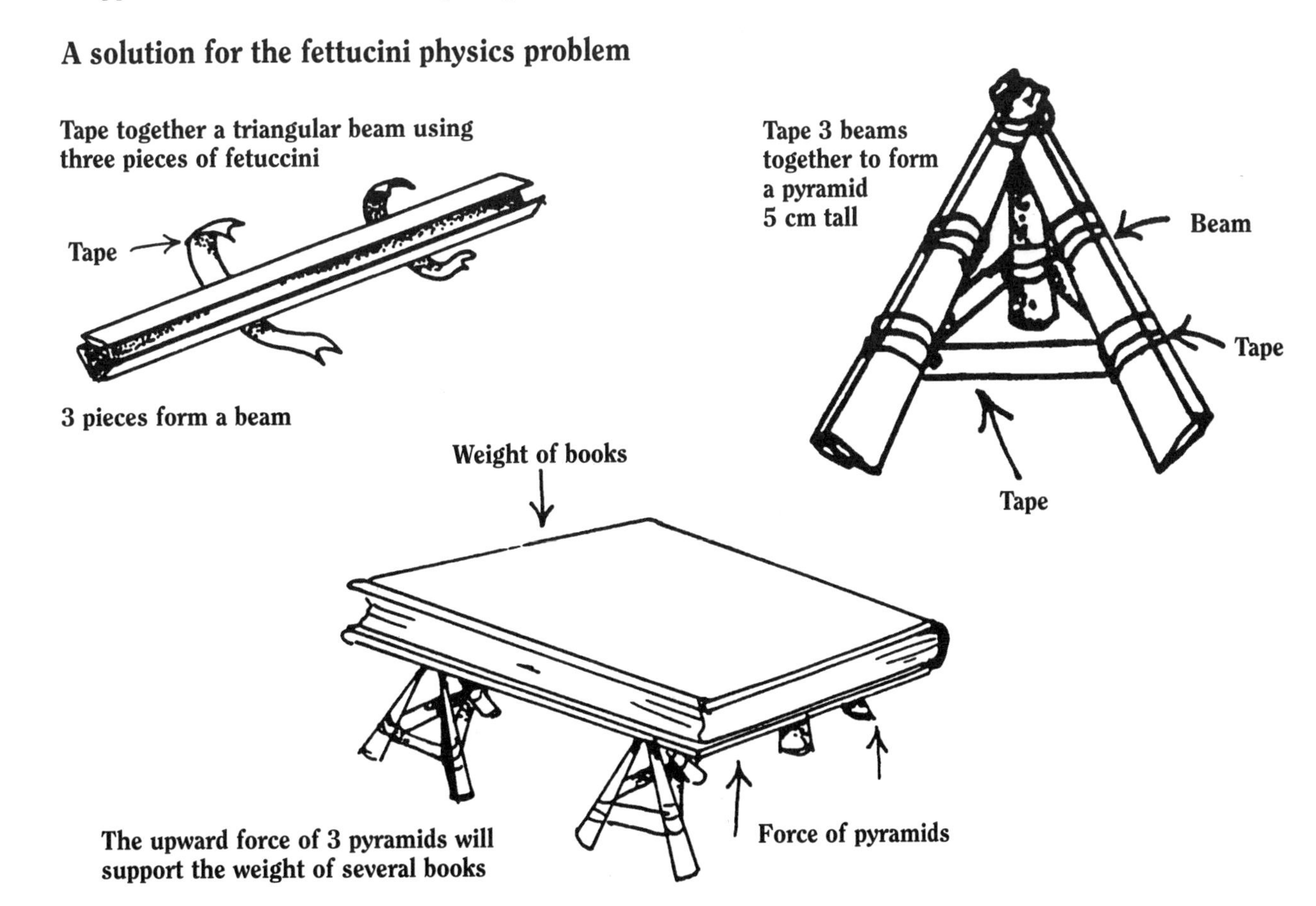

ACTIVITY 22 WORKSHEET

Hair Drier versus Gravity—Equal and Opposite Forces?

Materials

Each group will need
- a meter stick
- an electric hair drier
- a ping-pong ball
- plasticine-clay type paper adhesive
- **optional:** a clamp to hold the hair drier in place

◆Background

Objects that have more mass are pulled by gravity with greater force than objects with less mass. In this activity you will see that a greater force is needed to balance the force due to gravity on more massive objects.

◆Objective

To see if the force of gravity acting on a ping-pong ball can be balanced by the force exerted on the ball by a hair drier

◆Procedure

1. Set the hair drier on the "High Speed" setting, point it straight up, and turn it on. (Set the drier on "No Heat" if possible. The activity will work with hot air, but the drier nozzle may get hot enough to cause burns.)

! Be careful when handling a hair drier that has been operating on the heat setting. The nozzle can get hot enough to cause burns.

2. Hold the ping-pong ball about 25 cm above the nozzle in the center of the column of air coming from the drier. Gently release it. The ball should float in the air directly over the drier. If it flies out of the column of air, try releasing it from different heights until you find a release point that allows it to continue floating.

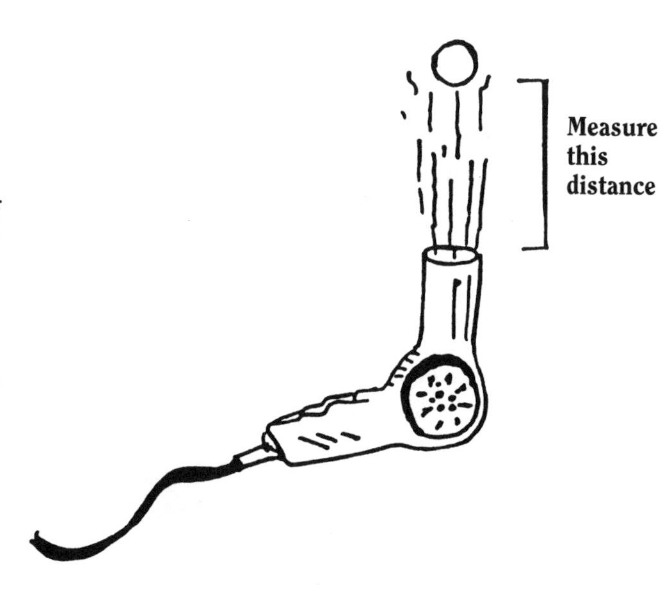

! Be sure the adhesive balls are firmly stuck to the ping-pong ball. The adhesive may damage the drier if it falls into the nozzle.

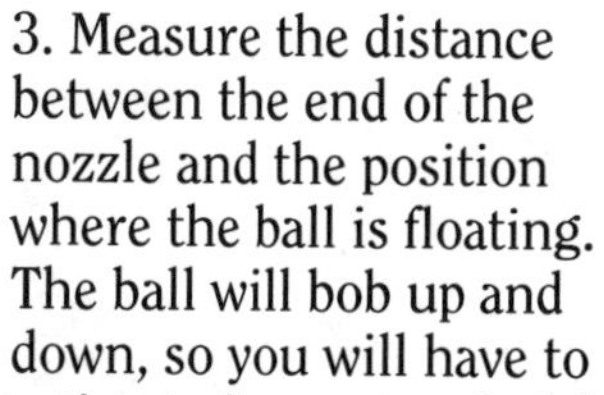

3. Measure the distance between the end of the nozzle and the position where the ball is floating. The ball will bob up and down, so you will have to estimate its average height to the nearest centimeter. Record your measurement in the data table.

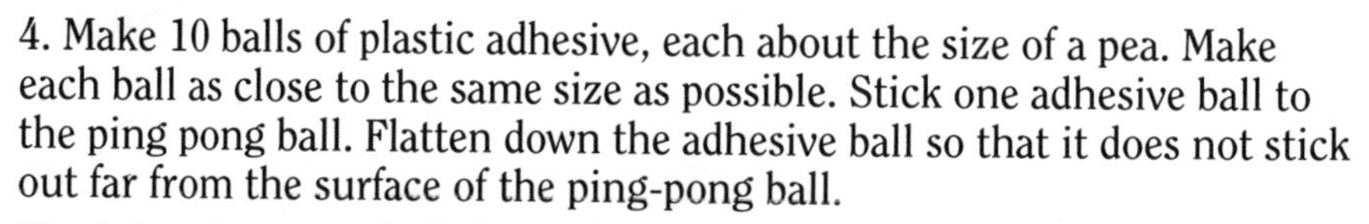

4. Make 10 balls of plastic adhesive, each about the size of a pea. Make each ball as close to the same size as possible. Stick one adhesive ball to the ping pong ball. Flatten down the adhesive ball so that it does not stick out far from the surface of the ping-pong ball.

Float the ping-pong ball above the drier and measure its average height above the nozzle as before. Record your measurement in the data table.

5. Stick another adhesive ball to the ping-pong ball, float the ping-pong ball above the drier, and measure the height as you did for steps 3 and 4. Record your measurement in the data table.

Continue adding adhesive balls one at a time and measuring the height until the ping-pong ball will no longer float above the drier.

Data table

Adhesive balls on ping-pong ball	Approximate height (cm)
0	_______ cm
1	_______ cm
2	_______ cm
3	_______ cm
4	_______ cm
5	_______ cm
6	_______ cm
7	_______ cm
8	_______ cm
9	_______ cm
10	_______ cm

6. Hold your hand directly in front of the nozzle of the drier. Move your hand to within 1 cm of the nozzle and notice how hard the air pushes against your hand.

Now move your hand 20 cm from the nozzle and notice the push from the air; move 1 m away; move 5 m away.

Describe the relationship between the distance from the nozzle and the push (amount of force) exerted by the air from the drier.

7. What is the force that pulls the ping-pong ball toward the center of the Earth?

8. What is the force that pushes the ball away from the Earth?

9. The forces you named for questions 7 and 8 act in *opposite directions*. Is there evidence that these opposite forces are *equally strong*? Explain.

10. Sticking the adhesive balls to the ping-pong ball increases its weight. Summarize the general relationship between the *weight* of the ping-pong ball and the *height* of the ball above the nozzle.

GUIDE TO ACTIVITY 22

Hair Drier versus Gravity—Equal and Opposite Forces?

◆What is happening?

The air coming from the hair drier exerts a force on any object that it strikes. When placing your hand in front of the drier, you can sense that the force is strongest close to the nozzle, and becomes progressively weaker as you move away from it. The magnitude of the force that the air exerts decreases as the distance from the nozzle increases.

The change in magnitude of the force is caused by collisions between the gas molecules making up the air in the room and the gas molecules being propelled by the drier. The air shot out of the drier crashes into the room air and is slowed down. The farther a fast-moving molecule travels away from the nozzle of the drier, the more likely it is to be slowed down by colliding with a slow-moving molecule in the room air. Another reason the force decreases as you move away from the nozzle is that the air flow expands as it leaves the nozzle. The farther away you are from the nozzle, the less dense will be the air with the fast-moving molecules.

Since the drier is pointed straight up, the force it exerts on the ping-pong ball is *opposed by* the downward force of the weight of the ball. When the upward force exerted by the air from the hair drier is *equal* to the downward gravitational force on the ball, the net force on the ball is zero. By Newton's second law, $F = ma$, if the force is zero and the mass is not zero, then the acceleration is zero. So there is no change in motion and the ball floats in place.(We can disregard its bobbing, which is caused by irregularities in the air flow.)

Sticking plastic adhesive on the ping-pong ball increases the downward force that gravity exerts on the ball. The downward force is now *unequal* to the opposing upward force of the air from the drier, so the ball accelerates downward (the force is no longer zero, so the acceleration cannot be zero).

As the ball approaches the nozzle, the force of air from the drier becomes stronger. The ball reaches a level in the column of air where the upward force of the air is equal to the downward force of gravity. The ball comes to rest at that height since the net force, and therefore the acceleration, is once again zero.

◆Time management

One class period (40–60 minutes) should be enough time to complete the activity and discuss the results.

◆Preparation

This activity requires a "heavy duty" or "pro" hair drier that blows a large volume of air through its nozzle. Small driers do not put out enough air to support the ping-pong ball high enough to allow multiple measurements to be performed. The wattage listed on the drier is *not* a reliable indication of the air output—test the driers before beginning this activity with a class.

Students may have some difficulty at first deciding on the true height of the ball. The measurement obtained will be an eyeball average of the position of the ball. You may wish to use this as another opportunity to discuss the fact that some error in measurement is unavoidable, no matter how refined the technique of measurement being used.

Flattening the adhesive on the ping-pong ball prevents an interesting, but somewhat puzzling result: if the adhesive pieces are small and stick out far enough from the sides of the ball, the ball may (initially, at least) *float higher* as additional pieces of adhesive are stuck to it. The adhesive increases the surface area of the ball, and the force of the air acting on the larger surface can, in some cases, give the ball additional lift.

Eberhard Faber's Holdit™ (or similar products for hanging posters on walls) is the preferred material for adding weight to the ball. These adhesives are generally available in office supply stores or housewares departments of drug and department stores. Moist, well-chewed bubble gum may be substituted if Holdit is not available.

If you have difficulty obtaining a sufficient number of powerful hair driers to perform this activity as a student investigation, the following substitutions are recommended:

- If one hair drier is available, perform the activity as a demonstration with assistance from students.
- Use alternative air sources: the blowing end of a vacuum cleaner works well for this activity.
- A similar demonstration can be done using a fan and a balloon or beach ball. Many fans have different speed settings, so the force of the fan can be varied.

◆Suggestions for further study

Sticking adhesive on the outside of the ball degrades its aerodynamic properties somewhat. If you have students who want to try a more elegant approach, have them add mass to the ball by injecting small amounts of glue into the ball and letting it harden. The ball will be unbalanced, but the surface will remain smooth.

◆Answers

3–5. *The heights of the ball will vary*, depending on the power of the hair drier and size of plastic adhesive balls used. The following heights are included only as a general guide to the possible results: 0 adhesive balls, 26 cm; 1 adhesive ball, 23 cm; 2 adhesive balls, 19 cm; 3 adhesive balls, 16 cm; 4 adhesive balls, 14 cm; 5 adhesive balls, 10 cm. More than 5 adhesive balls could not be supported by the force of the air coming from the drier.

6. The force exerted by the drier decreases as you move away from the nozzle. From 5 m away, it is difficult to feel the air stream at all. Close to the nozzle, the force is quite strong.

7. Gravity pulls the ball toward the center of the Earth.

8. The air coming from the drier pushes the ball away from the Earth.

9. At the height where the ball floats steadily, the upward force of the air from the drier equals the downward gravitational force on the ball. At that level, the ball is neither falling nor rising (except for its small bobbing motions caused by irregularities in the air flow.)

10. The greater the weight of the ball, the closer to the nozzle it floats.

MODULE 6

"Hidden Forces" Affecting Motion

◆Introduction

• How can a baseball pitcher's curveball curve? The first law of motion predicts straight-line motion.

• Why does a balloon full of air fall more slowly than a rock?

• Astronauts orbiting the Earth experience weightlessness. Have the astronauts lost *mass*? Have they completely escaped the effects of gravity?

Our observations of objects moving near the surface of the Earth often seem to contradict Newton's laws of motion. The first law predicts straight-line, constant velocity motions, yet our experience tells us that such motion rarely seems to occur in real life.

The activities in Module 6 allow you to identify and measure some of the "hidden forces" that affect motion; these forces are hidden in the sense that their presence is sometimes easily overlooked, but recognizing them is essential to describing and predicting the motion of objects on Earth. You will discover ways of reducing or eliminating the effects of these forces. Films made by astronauts during a Space Shuttle mission allow you to see what happens when gravity, one of the pervasive unequal forces, is canceled out.

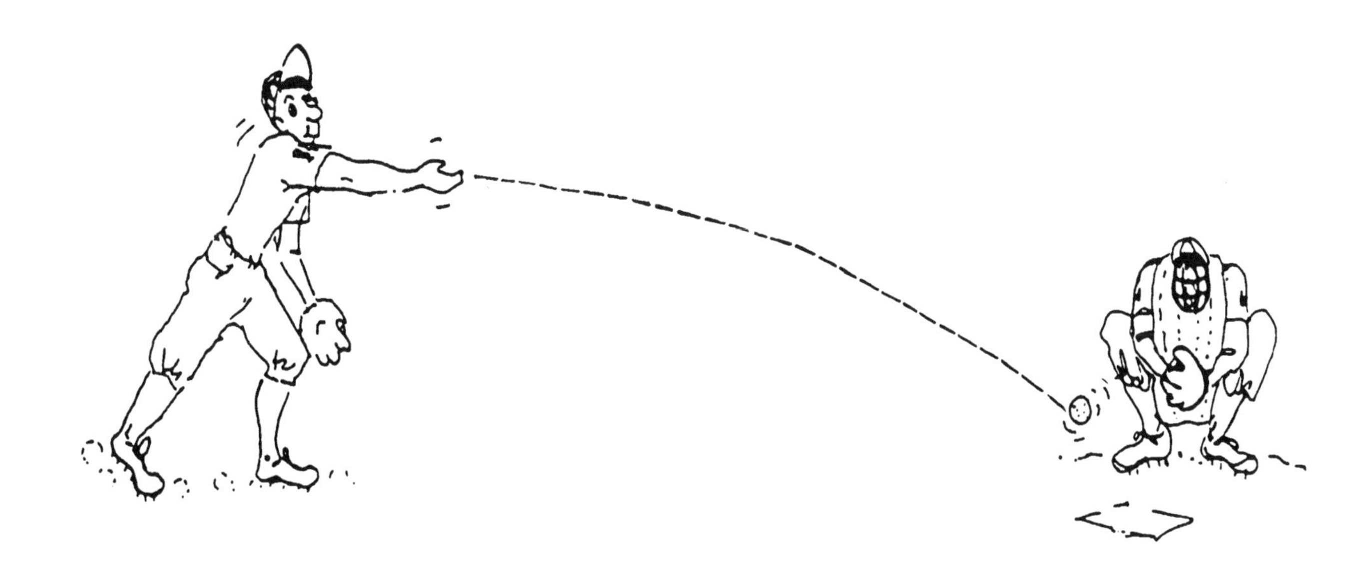

◆Instructional Objectives

After completing the activities and readings for Module 6, you should be able to

- identify forces that prevent most objects near Earth from traveling in a straight line at a constant speed [Activity 24 and Reading 13]
- measure the frictional force affecting a sliding object [Activity 23]
- demonstrate the effects of air resistance (drag) on falling objects [Activity 24]
- distinguish between weight and mass [Activity 26]
- explain why astronauts in orbit are weightless [Activity 25 and Reading 14]

◆Preparation

Study the following readings for Module 6:

Reading 13: Can Objects "Break the Laws of Motion?"

Reading 14: Gravity, Weight, and Weightlessness

◆Activities

This module includes the following activities:

Activity 23: Measuring the Force Required to Move a Wooden Block

Activity 24: How Do "Hidden Forces" Affect Falling Objects?

Activity 25: *Eureka!* #6— Gravity

Activity 26: *Eureka!* #7— Weight versus Mass

Activity 27: Is Gravity Essential?

ACTIVITY 23 WORKSHEET

The Force of Friction Acting on a Wooden Block

◆Background

The force of **friction** is one of the "hidden forces" that affect the motion of objects on Earth. Forces like these are hidden in the sense that they are so pervasive that we sometimes take them for granted and overlook them. The force due to friction and other hidden forces must be recognized if we are to describe and predict motion accurately.

◆Objective

To measure the force due to friction acting on a piece of wood

◆Procedure

1. Adjust the spring scale so that it reads zero. Use the cup hook to attach the spring scale to one end of the board as shown in the diagram. Place the scale and board on a smooth, level surface.

Materials

Each group will need

- a 35-cm length of lumber (a piece of 2 x 4 board)
- a cup hook
- spring scale calibrated in newtons (or gram scales may be used)
- 500-g mass (a 1-pound can of soup, lead sinkers totaling 16 oz, or other object of similar mass may be used)
- 4 or more *round* pencils or sections of dowel

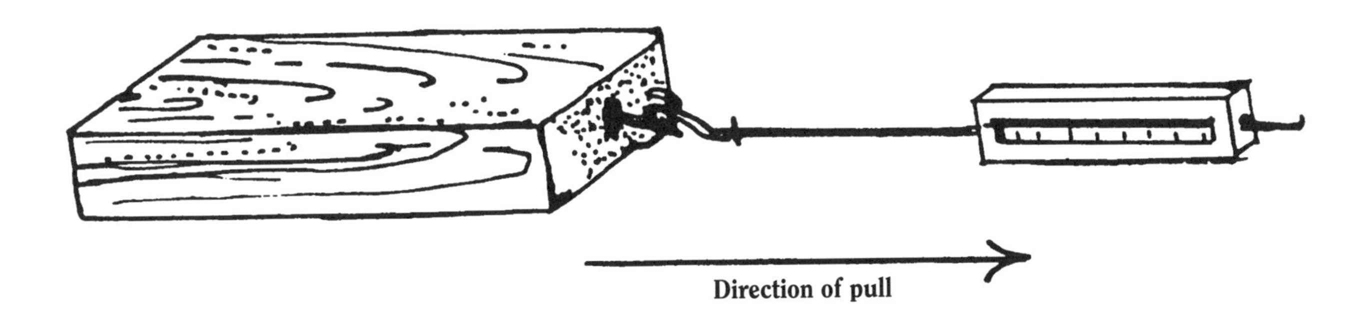

2. Gently pull on the scale so that the board moves in a *straight line* and at a *constant speed.*

Read the scale *while the board is moving*. In the data table below, record the number of newtons required to keep it moving at a constant speed. Repeat the reading three times.

Vocabulary

- **Friction:** Resistance to motion between two bodies in contact.

Data table

Force required to pull board

	Trial 1	Trial 2	Trial 3	Average
Board alone	_____ N	_____ N	_____ N	_____ N
Board + weights	_____ N	_____ N	_____ N	_____ N
Board + weights on rollers	_____ N	_____ N	_____ N	_____ N

3. Place masses totaling 500 g on the board. Pull the board exactly as you did before.

Read the scale while the board is moving *at a constant speed.*

Record on the data table the number of newtons required to move the board plus the weights. Repeat the reading three times.

4. Place the pencils under the board to serve as rollers. With the weights still on the board, pull the board at the same speed as before.

Record on the data table the number of newtons required to roll the board plus the weights.

5. Which of the three conditions you tested (board alone; board plus the weights; or board plus the weights on rollers) required the *greatest* force to move the board at a constant speed?

____

6. How do the rollers affect the force required to move the board?

____

____

____

____

7. Describe how the reading on the scale changes when the board is set in motion.

____

____

____

____

8. Newton's first law of motion states "...an object will remain in motion in a straight line and a constant speed *unless* acted upon by an unequal force." The block stops moving soon after you stop pulling. What is the unequal force (or the forces) that change the block's motion?

____

____

____

____

GUIDE TO ACTIVITY 23

The Force of Friction Acting on a Wooden Block

◆What is happening?

The force being measured results from the **friction** between the board and the surface it slides across. The force due to friction is one of the important "hidden forces" (forces that are often overlooked) that modify the motion of all objects on Earth.

As mass is added to the board, the downward force that the board exerts on the surface *increases* and the friction between the board and the surface *increases*. The additional mass also increases the inertia or "difficult-to-moveness" of the board. More force is required to move the loaded board.

The rollers (pencils, dowels, or even empty soup cans) placed under the board *decrease* the friction between the board and the surface. Therefore, less force is required to move the board when it is on rollers.

Careful observers will note that *just before* the board starts to move, the spring scale reaches its maximum reading. The reading *decreases* after the board is moving at a constant rate. The differences in readings are caused by several factors:

- Force is required to accelerate an object from zero to a constant speed. Initially, the scale registers the force required to overcome the static friction.
- Force must be applied in order to overcome the frictional force acting on the board. If there were no friction, after a block was set in motion it could be released and it would continue coasting. No additional force would be required to maintain its motion, and the scale would read zero.
- Friction is less between a moving object and a surface than it is between a non-moving object and a surface. Therefore, less force is required to overcome the friction acting on the board moving at a constant speed than is required to break it loose and start it moving.

◆Time management

One class period (40–60 minutes) should be enough time to complete the activity and discuss the results.

◆Preparation

If you do not have spring scales calibrated in newtons to use in class, scales calibrated in grams may be used. The use of gram scales is discussed in the "Preparation" section of Activity 7.

If no spring scales are available, a simple force measurer can be made by taping a rubber band to one end of a ruler. Tie the board to the opposite end of the rubber band, and use the ruler to see how far the rubber band stretches when you pull on the board. The stretch of the rubber band increases as the force increases. You might also substitute any small hand-held scale (such as a fisherman's scale) to determine force.

Empty cans may be substituted for the pencils or dowels used as rollers. Be sure that there are no jagged edges on the cans. If cuphooks are in short supply, you may wish to use masking tape to attach the string to the board.

◆Suggestions for further study

Have students devise and test other methods of reducing the friction between the board and the table, such as using hexagonal pencils to support the board, lubricating the board, or placing the board on wax paper.

Suppose you wanted to make the fastest trip possible down a snow-covered hill. Would you choose to use a toboggan with a large bottom surface or a pair of skis with a small bottom surface? Which would result in the greatest friction? Have students use the wooden block to find an answer to this question. Suggest they think of the large surface of the block as the toboggan and the small surface as the skis.

◆Answers

2–4. *The exact numerical values for observations will vary* depending on the surface of the table, the roughness of the board, and the amount of weight used. However, the expected trends are as follows:

The force (in newtons) required to move the board will *increase* when mass is added to the board.

The board plus the weights placed on rollers will require *less* force to move than the board plus the weights sliding on the table.

5. The board plus the weights require the most force to keep moving at a constant speed.

6. The rollers decrease the force required by decreasing the force of friction between the board and the table top.

7. The scale gives a very high reading just before the board first starts moving. The reading then goes down and stays at a lower reading as long as the board is moving at a steady speed.

8. Friction between the board and the surface it slides along acts as an unequal force that stops the board's motion. Air resistance (drag) is also an unequal force, but it is much smaller than the surface friction.

ACTIVITY 24 WORKSHEET

How Do "Hidden Forces" Affect Falling Objects?

◆Background

The question "Do heavy objects fall faster than light objects?" cannot be answered completely until we consider another of the "hidden forces"—**air resistance.** Air resistance is like other hidden forces in that it can be easily overlooked. In this activity we will demonstrate the existence of air resistance.

◆Objective

To demonstrate how air resistance affects the fall of objects toward the Earth

Materials

Each group will need

- a meter stick
- a timer which can measure 0.1-s intervals
- a 2-cm length of yarn
- a penny
- a hardcover book
- 10 cm of masking tape

◆Procedure

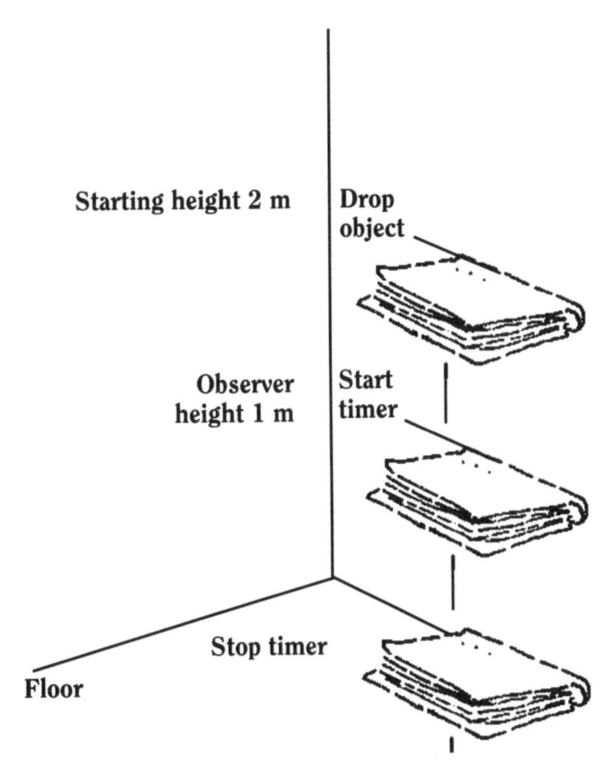

Vocabulary

- **Air resistance:** A force exerted on a moving object opposite to its direction of motion due to the friction between the object and air. Air resistance is also called *drag* or *air friction.*

1. Place a small piece of masking tape on the wall 2 m above the floor to mark the starting height for all objects.
Mark a spot 1 m above the floor with another piece of masking tape. The time observer should stand with his or her eyes level with this mark, holding the timer.

2. Hold the penny even with the 2-m mark.
When the time observer is ready to start the timer, drop the penny.

3. When the penny reaches the 1-m level, the time observer starts the timer.
Stop the timer when the penny hits the floor.

Hint: *This activity requires you to measure very short time intervals. Practicing steps 2 and 3 several times before recording your observations will improve the accuracy of your timing.*

4. Record in the data table how much time the penny required to fall the second meter.

Repeat your time measurement three times for each object listed in the table below.

Data table

Seconds to fall final 1 meter	Trial 1	Trial 2	Trial 3	Average time
penny	_____ s	_____ s	_____ s	_____ s
yarn	_____ s	_____ s	_____ s	_____ s
book*	_____ s	_____ s	_____ s	_____ s

*drop the book so that it *lands flat* on its cover!

5. Calculate the average speed that each object was moving as it fell the *second meter*. State your results in meters per second.

penny _____ m/s yarn _____ m/s book _____ m/s

6. Why is it difficult to measure accurately the time interval (Δt) that an object requires to fall the second meter?

____

____

7. Write a statement that summarizes your observations about the rate at which the different objects fell.

____

____

8. What hidden force caused the yarn to take longer than the book or penny to fall the final meter?

____

____

9. Place the yarn and the penny on the top cover of the book. Hold the book 2 m above the floor and drop it so that it *lands flat* on the bottom cover. Repeat this procedure 3 times.

Do all three objects fall at the same rate? Describe your observations.

____

____

10. Can you explain why the book affects the rate at which the yarn falls?

____

____

11. Do heavy objects always fall to Earth faster than light objects? Why or why not?

____

____

GUIDE TO ACTIVITY 24

How Do "Hidden Forces" Affect Falling Objects?

◆What is happening?

This activity is designed to demonstrate the existence of an invisible force, air resistance, that affects how objects fall. When asked if heavy objects fall faster than light objects, many people would answer "yes." Sometime before 322 B.C. Aristotle stated that heavy objects fall faster than light objects. His pronouncement was accepted as fact for almost 2000 years. But was his understanding of motion complete?

In this activity, comparing the rate of fall of the penny with the rate of fall of the yarn seems to support Aristotle's view. But why does the book, which is the heaviest object tested, seem to fall at about the same rate as the penny? Additional experimentation leads to a different (and correct) conclusion: the force of friction between the air and a falling object affects an object's rate of fall.

In the first two meters of fall, the book and the penny are affected very little by the frictional force of the air. Both objects accelerate toward Earth at about 10 m/s^2 when dropped. The friction between the air and these objects increases as their speed increases. However, the book and penny continue to accelerate because the *upward* force of air resistance acting on them is much smaller than the *downward* force of gravity.

The effect of air resistance acting on the yarn is easier to observe. The yarn requires more time than the penny or book to fall the final meter. Unlike the penny and book, the yarn has a large amount of surface area per unit of weight. The larger the surface area per unit weight of a falling object, the larger will be the upward force of air resistance, at the same speed. Since the weight of the yarn is small, air resistance measurably decreases the yarn's rate of descent. The yarn acts as its own parachute.

Initially, the yarn accelerates at 10 m/s^2, just like the penny and the book. As its speed increases, air resistance (the friction between the air and the yarn) increases. Because of its shape and light weight, the upward force of air resistance becomes *equal to* the downward force of gravity before the yarn falls one meter. When these forces become equal, the *yarn stops accelerating* (in $F=ma$, $F=0$ means $a=0$), but it continues to fall at a constant speed.

In 1638, Galileo challenged Aristotle's view by publishing the hypothesis that "in a vacuum, all bodies would fall with the same speed." It is questionable whether he reached this conclusion by actually performing the famous experiment of dropping objects from the Leaning Tower of Pisa, but his prediction is correct. In the absence of the invisible force of air resistance, the weight of an object *does not* affect its rate of falling toward Earth.

Can air resistance be reduced?

Dropping the yarn and penny on top of the book shows one way that air resistance can be reduced without placing both objects in a vacuum chamber. By pushing some of the air out of the path of the yarn, the book decreases the force that the air exerts on the yarn. As a result, the yarn falls at about the same rate as the book and penny.

If this experiment were performed in a vacuum, there would be no frictional force caused by the air. All three objects would fall at exactly the same rate.

Sources of experimental error

Even without actually timing the falling objects, most observers will agree that the penny and book fall faster than the yarn. However, making an accurate *quantitative* determination of the differences among the rates of falling for the three objects (by timing the final meter of fall) is not easy. Differences in the observers' reaction times when starting and stopping the timer will produce a substantial amount of variation in time measurements.

The difficulty of timing the final meter of fall can be used as the basis for a discussion of the effects of measurement errors on the interpretation of experiments. Careful and comprehensive determination of potential sources of error in an experimental design is an important scientific process. All experiments involve some degree of error that cannot be eliminated; identifying errors and not allowing them to lead to incorrect conclusions is essential to scientific inquiry.

◆Time management

One class period (40–60 minutes) should be enough time to complete the activity and discuss the results.

◆Preparation

Timing errors are certain to occur when students are performing this activity. You may wish to discuss how timing errors may be reduced. In a recent Winter Olympics skiing event, a time difference of only 1/100 s separated the gold medal and silver medal winners. The timing was done by an electric eye (a photoelectric gate). In all previous Olympiads, the two skiers would have both been awarded gold medals, because the event was hand-timed to only 1/10-s intervals.

◆Suggestions for further study

The objects listed to be dropped for this activity are only suggestions—any similar group of objects may be used. You may wish to drop objects from a greater height to make differences in their rates of fall easier to observe.

◆Answers

The following data table is *for reference only*; answers will vary. *There is no one correct set of answers.*

Sample data

Seconds to fall final 1 meter

	Average time
penny	0.2 s
yarn	0.6 s
book	0.2 s

5. Answers will vary; expected values are about 5 m/s for the penny and book, and 1.5 m/s for the yarn. A sample calculation follows:
If $\Delta t \approx 0.2$ s and $\Delta x = 1$ meter, then the
average speed = 1 m / 0.2 s = 5 m/s.

6. Answers will vary; a statement that the interval is so short that the observer's reflexes aren't fast enough to work the timer accurately is hoped for. This question can be used to start a discussion of the limits of accuracy in measurement for different types of experimental designs.

7. The book and penny required about the same time to fall the last meter, but the yarn took longer. Some students may say that the heavier objects fell faster. This statement is true, but is an *incomplete analysis*: it ignores the force of air resistance. Air resistance causes the differences in the rate of falling.

8. The hidden force is air resistance or friction or drag.

9. The book, penny, and yarn all fall at the same rate and all hit the floor at the same time. With the book leading the way, the yarn falls at the same rate as the more massive objects.

10. The book reduces the frictional force of air resistance acting on the yarn. Therefore, when it is behind the book, the yarn falls at essentially the same rate as the book and the penny. If this experiment could be done in a vacuum chamber, all of the objects would fall at the same rate, since there would be no air resistance.

11. If the friction caused by air hitting the light object is reduced, light objects fall at about the same rate as heavy objects. The important criterion here is surface area per unit weight. Unless special efforts are made to reduce air resistance, however, light objects will generally fall more slowly than heavier objects.

ACTIVITY 25: VIDEOTAPE

Eureka! #6—Gravity

◆Background

Gravitation is one of the fundamental forces of the universe. All matter attracts (and is attracted to) other matter in the universe. Sir Isaac Newton's statement of the law of gravitation helped to unify the astronomical observations of Galileo, Brahe, and Kepler, and was the final step in the new science of **dynamics**.

◆Time management

The running time of the videotape is 5 minutes. At least 15 minutes should be allotted to introduce, run, and discuss the videotape. You may wish to play the videotape at the end of a lesson to reinforce the concepts presented.

◆Comments on the videotape

Concept 1 is a non-numerical statement of Newton's law of gravitation. The law is illustrated in two ways:

• by using vectors to represent the equal and opposite forces being exerted by gravity and the stem of an apple (while the apple is hanging from a tree)

• by showing that the Earth attracts apples, acrobats, oil paintings, and three green bottles with a force that we call gravity.

If there is no opposite and equal force (such as the acrobat exerts by holding onto the trapeze) opposing the force of gravity, all the objects will accelerate toward Earth.

Concept 2, relating mass to gravitation, is not illustrated graphically, and must be accepted "on faith" (in Newton). Gravitation is quite literally a "force of nature" that exists everywhere that there is mass.

Concept 3 relates the force of gravity to the effect that it produces: accelerating objects near the surface of the Earth at a constant rate of about 10 m/s^2. Newton's second law, F = ma, can be stated for the force of gravity as F = m*g*, where *g* represents the acceleration due to gravity and has an approximate value of 10 m/s^2. On planets having a different mass than Earth, the value for *g* would be different.

Concept 4 is a review of the standard unit of force, the newton. The newton is related to the acceleration of an average-sized apple having a mass of 100 g. A gravitational force of 1 N is exerted on the apple when it accelerates at a rate of 10 m/s^2. (Gravity accelerates *all* objects near Earth at a rate of 10 m/s^2 *regardless of their mass*).

This illustration of the newton shows that an *equal and opposite force* of 1 N (supplied by a hand) is required to prevent the downward fall of the apple. The apple exerts a force of 1 N on the hand, and the hand exerts a force of 1 N on the apple.

Concept summary

1. "All masses attract all other masses to a certain extent."*
2. "The Earth, being a very large mass, exerts a very large force of attraction which is called the force of **gravity**."*
3. "This pulls all objects toward Earth with an acceleration of roughly 10 meters per second per second."*
4. "The force of gravity acting on a mass of 100 grams is called 1 newton."*

Vocabulary

• **Dynamics:** A branch of mechanics that concentrates on the study of bodies in motion.

**Eureka!* Produced by TVOntario © 1981.

ACTIVITY 26: VIDEOTAPE

Eureka! #7—Weight versus Mass

◆Background

This segment of *Eureka!* begins with an illustration of several items that weigh about 1 N: an apple, a pair of golf balls, and one flashlight battery. The narrator points out that each of these objects weighing 1 N also has a mass of 100 g, and poses this question: Why bother with the distinction between mass and weight?

◆Time management

The running time of the videotape is 5 minutes. At least 15 minutes should be allotted to introduce, run, and discuss the videotape. You may wish to play the videotape at the end of a lesson to reinforce the concepts presented.

◆Comments on the videotape

In everyday life, most people (including the majority of scientists) do not need to worry about the formal differences between mass and weight. We may *incorrectly* say that an object *weighs* 20 kg, when we really mean that it *has a mass of* 20 kg, and weighs 200 N.

Until we leave the Earth and go to another planet, the distinction between mass and weight is, for most of us, unimportant. Physicists, however, need to be precise in differentiating between weight and mass, because they *are* concerned with the behavior of matter in places other than the 1 g environment near the surface of the Earth.

For example: knowing the gravity on the Moon, and understanding how it affects the weight of objects landing on the Moon, was crucial to the planning for the lunar voyages of Project Apollo. Scientists had to calculate very precisely how much force would be required for the lunar module to escape the gravity of the Moon. The astronauts might not have been able to return to Earth if the calculations of the moon's gravity had been incorrect.

Knowing that the cargo (astronauts and their research equipment) being carried across the moon's surface by the Moon buggy would only weigh 1/6 as much on the Moon as they did on the Earth allowed the engineers to design a very lightweight vehicle. In the 1 g environment of Earth, the Moon buggy might have been too fragile to carry the load for which it was designed; the reduced gravity of the Moon allowed it to function efficiently.

After viewing this segment of *Eureka!*, you may wish to review the three definitions of mass given in Module 1. Concept 1 in this segment of *Eureka!* restates one of the definitions; the other two (in abbreviated form) are:

- mass is a measure of an object's resistance to change of motion
- mass can be determined using a two-pan balance

Concepts 2 and 3 of this program relate the concept of weight to the observable phenomenon of acceleration due to gravity. Without stating the equation $F = mg$, the middle segment of the tape shows that the weight of an object depends on local conditions—specifically, the amount of gravitational force acting on the object. Gravitational force is determined by the total amount of mass present in a system.

The implications of Concept 4 are amusingly illustrated with the

Concept summary

1. "Mass refers to how much 'stuff' a thing contains."*
2. "Weight refers to the rate at which that amount of stuff is being accelerated towards the Earth."*
3. "In other words, weight is just another way of saying *force of gravity*."*
4. "An object's mass never varies, but its weight can go up or down, depending on the force of gravity acting on it in various parts of the universe."*

example of a slightly rotund astronaut. The narrator comments that if you are not slim, you should not eat as much. You may need to join "Masswatchers." But if you just want to lose weight, go to the Moon. Your body will consist of just as much "stuff," but your scale will tell you that you weigh *much* less. Gravity causes weight to vary; changing gravity does not affect mass.

ACTIVITY 27 WORKSHEET

Is Gravity Essential?

◆Background

As long as we are on Earth we cannot escape the force of gravity. We are so accustomed to gravity that it is difficult to imagine life without it. In this activity your teacher will demonstrate the way some simple toys work at 1 *g* (the force of gravity on Earth). You will try to answer questions about how the toys would work at 0 *g* (weightlessness) aboard an orbiting spacecraft. You will then watch a videotape of toys in space to see how good your predictions were.

Materials

Each group will need

- a paper airplane
- a Slinky™
- a yo-yo
- a set of jacks and a ball
- a Whee-lo™
- *Toys in Space* videotape

◆Objective

To predict how some simple toys would work in the absence of gravity

◆Procedure

1. Paper airplanes glide across the room in a downward sloping path at 1 *g*. Describe the path an airplane would travel at 0 *g*. Would it glide up, glide down, or go straight?

____

2. The Slinky sags in the middle into a curve when it is held up by its ends at 1 *g*. Would a Slinky sag downward at 0 *g*? Why?

____

____

____

____

3. Yo-yos fall downward toward the center of the Earth when they are released at 1 *g*. The string unwinds. When the yo-yo reaches the end of the string, the string rewinds and the yo-yo climbs the string. At 0 *g*, the yo-yo would not fall and unwind the string. How could you make it work? Would it return?

____

____

____

____

Hint: *Similar questions stumped a group of astronauts before a space shuttle mission. Do not worry about whether your answers are right or wrong!! Just try to answer the questions about the toys by applying what you know about gravity and the laws of motion.*

4.To play jacks at 1 *g*, you toss a ball, pick up some jacks, and catch the ball after it bounces once. At 0 *g*, the jacks and ball would float. How might you change the rules to make it possible to play jacks in space?

5. A Whee-lo can be set in motion by tilting the metal tracks downward. The wheel rolls around the inside and the outside of the tracks as you tip it back and forth. A magnet holds the wheel on the tracks. How could you make the wheel move at 0 *g*?

6. View the *Toys in Space* videotape and see how the toys behave in a weightless environment. How good were your predictions?

GUIDE TO ACTIVITY 27

Is Gravity Essential?

◆What is happening?

Don't be discouraged if your intuition about how the toys would work is proven wrong. The astronauts found some results that surprised them, too.

Films of objects moving aboard orbiting spacecrafts are the only means by which most people can see objects moving in a straight line exactly as Newton predicted in his laws of motion. All objects obey the laws of motion in the 1 g environment of Earth as well as in space. However, accounting for all the "hidden forces" (those forces, such as gravity and air resistance, that are so pervasive that we sometimes take them for granted and overlook their presence) that combine to produce curving paths or acceleration is sometimes difficult.

Newton's synthesis of the laws of motion seems even more remarkable when one considers that he did not have access to a weightless environment. He was somehow able to extract the elegant simplicity of straight-line, constant velocity motion from the midst of the confusing, curving, accelerating motions of objects he observed near Earth.

One of the interesting things that the narrator in the film points out is that gravity is so pervasive a force that we often fail to recognize all of its effects. Even the scientists among the space shuttle crew were not able to predict exactly how simple toys such as a yo-yo would perform under weightless conditions.

Many of the experimental procedures included in these six modules in Newtonian mechanics are designed in such a way that the effects of gravity are in some way canceled out by an equal and opposite force. We can never truly escape the effects of gravity, however, unless we go into orbit as these astronauts did on Space Shuttle Mission STS 51D.

Both scientists and science teachers sometimes think that science experiments must be complicated, *very serious* endeavors using equipment from manufacturers of scientific apparatus if valid results are to be obtained. However, the experiments shown in this tape were performed with toys costing only a few dollars each; the logic of the experimental designs was very simple and the experimenters obviously enjoyed themselves thoroughly while doing the work. We cannot use the very expensive and complex space shuttle as a laboratory as did these astronauts, but we can emulate their willingness to perform experiments using simple devices such as tops, jacks, and wind-up mice.

Many of us would like to "leap tall buildings in a single bound" like Superman. In the weightless conditions of space, we could exert enough force to make those leaps. But without *some* gravity (or a comparable force) to hold us down against a suitable surface, the action of suddenly straightening out our legs would not cause the desired reaction of clearing the top of the building. The would-be Superman's legs would simply flail against the air without propelling him forward at all.

Objects released in the gravity-free environment of the space shuttle float away from the point of release. Tiny forces that would be too small to overcome Earth's gravitational force (such as a slight push by a finger while releasing "Rat Stuff") are sufficient to move objects in space.

Space Shuttle Pilot Donald Williams discovered that Rat Stuff the mouse could not do his trick (flipping and landing on his feet) because there was insufficient gravitational force to hold the mouse on a surface.

The small unequal forces exerted by Williams while releasing the toy caused it to float away from the bulkhead.

Williams applied hand cream to the toy's feet. (A small piece of Velcro™ eventually replaced the cream.) Once Rat Stuff's feet could push (exert a force) against something solid, he sailed straight across the cabin. Although he rotated about his center of mass, he could not perform a flip and land on his feet. The flip trick requires gravity to pull Rat Stuff back down. In the absence of gravity (which acts as an unbalanced force when Rat Stuff does his trick on Earth), the mouse exhibits straight-line motion as predicted by Newton's first law of motion.

Playing jacks in the weightless environment of the orbiting space shuttle was trickier than Dr. Rhea Seddon, Mission Specialist, anticipated. On Earth, gravity assists jacks players in ways that most people do not think about: it supplies the force to hold the jacks on a playing surface, it causes the ball to curve back to Earth so that the player can catch it, and it prevents the small action forces exerted by air currents from causing the reaction of dispersing the jacks in all directions.

Dr. Seddon found that weightlessness interfered with her timing for catching the ball. In the absence of the unequal force of gravity, the ball traveled in a straight line when she threw it. The ball's return depended on having it bounce off the walls of the cabin. The force that the ball applied to the wall was opposed by an equal and opposite force of the wall pushing back against the ball. The ball changed directions by 180°, and went back toward its point of release.

One might ask the question, "Since the ball is weightless, why does throwing it require the astronaut to exert a force with her hand?" The force required to throw the ball against the cabin wall can be explained using the second law of motion, F = ma. Throwing involves changing the ball's motion, or in other words, accelerating the ball.

A ball floating in the middle of the cabin is weightless; however, it possesses the same *mass* that it had on Earth before launching. *Weightlessness does not reduce mass.* One definition of mass is *resistance to change in motion* (inertia). Since the ball's mass is the same in orbit as it was on Earth, the amount of force required to overcome its resistance to change in motion is the same in space as on the Earth.

Playing jacks in a weightless spacecraft creates unexpected problems, but also has some major advantages. Since the jacks float in midair, there is no friction between the jacks and a playing surface. They therefore tend to fly in all directions when released, but the reduced friction also allows a jack to spin beautifully for long periods of time, slowed only by air resistance. The difficulty of doing "twelvesies" is certainly offset by the ease of performing the victory flip that Dr. Seddon demonstrated.

◆Time management

One class period (40–60 minutes) should be enough time for students to experiment with the toys, view the videotape, and discuss the results.

◆Preparation

Students will become much more involved with the film if they develop their own hypotheses about the outcome of the experiments being shown before seeing the film. One way of doing this is to have the students meet in small groups to see if they can agree or disagree with each other about what will happen. You may want to set up an "activity center" for each group of 4 to 5 students. Each activity center should have one set of all the toys needed, and each group can work out its predictions together. Have each group present their conclusions. Can the class reach a consensus about what will happen to each toy in space?

Demonstrate as many of the toys tested in space as possible before showing the film. The toys included on the tape are a paper airplane, a paddleball, a Slinky, ball and jacks, balls for juggling, magnetic marbles, a flipping wind-up gyroscope, a top, a Whee-lo, a yo-yo, and a toy car. Most of these (or similar items) are readily available at toy stores.

◆Suggestions for further study

Have students discuss what changes they would have to make in their daily lives if the Earth were a weightless environment. How would the absence of gravity affect eating? How would they go about brushing their teeth or sleeping at night? How would it affect traveling by car? Have students suggest ways that technology could compensate for these effects.

◆Answers

1. The paper airplane hangs motionless in the air until it is given a gentle push. It flies in a straight and level path until it hits the bulkhead. (**Note:** If the wings or tail of the airplane were uneven, or had flaps been built into it, the airplane would curve. The air passing over the irregular surfaces would create a force that would change its straight-line motion.)

2. The Slinky does not sag because there is no net downward force acting on it. This allows the astronauts to use the spring to demonstrate wave patterns without the distortion that gravity usually causes.

3. The yo-yo will not fall, but it can be flipped to the end of its string. It behaves normally for the most part except that it will not sleep (continue turning at the bottom of the string). Since it travels in straight lines when thrown at 0 *g*, certain tricks are much easier to perform.

4. Playing jacks at 0 *g* presents several problems: the jacks float freely, since there is no gravity to hold them to a surface, and the ball must be bounced off of something to get it to return to the player. A slight touch to the jacks, or even air currents, will cause them to disperse around the cabin. The trick to playing the game is keeping the jacks as motionless as possible in mid-air, and getting a smooth, predictable bounce of the ball off the bulkhead.

5. A Whee-lo can be operated at 0 *g* by slinging the wheel away from the handle. Once the wheel is rotating, the orientation of the handle does not matter. The wheel continues turning until friction between the wheel and the track causes the wheel to stop.

MECHANICS

Readings

◆Introduction

The following readings review and expand on the concepts introduced in the activities in Modules 1–6. Many teachers may wish to learn more about a particular topic than is included in these materials. *Conceptual Physics* by Paul G. Hewitt (Little, Brown and Company, 1985) is an excellent textbook to use for this purpose.

Dr. Hewitt encourages readers to develop an intuitive understanding of the everyday applications of physics; he does not stress "number crunching." The book's illustrations are both amusing and informative. Many chapters list home projects, simple experiments, demonstrations, and tricks for illustrating a particular concept.

READING 1

Measurement Skills Used in the Study of Moving Objects

◆Introduction

Scientific investigations of motion require direct measurements of mass, length, and time. While using certain mathematical concepts is helpful when describing motion, understanding how objects move does not necessarily require complicated mathematical analysis. The following sections provide some background information needed for studying motion.

◆Determining length

Length measurements are essential to scientists for computing speed, velocity, and acceleration.

The length of any object can be measured in units such as centimeters or meters. The standards for these units of length may consist of the distance between two marks on a metal bar, or they may be related to the wavelength of light or some other reproducible physical phenomenon.

Metric units (centimeters and meters) will be used for all length and distance measurements made for the activities in these modules. However, for practical reasons, instructions for building certain pieces of equipment used in the activities will include English (foot and inch) dimensions in cases when the materials being used (such as lumber) are commonly sold in non-metric units.

◆Determining time intervals

In order to keep up with the behavior of a chunk of matter in motion, scientists must measure intervals of time in some fashion. We generally define time by a convenient mechanical standard such as the motion of the hands on a clock, the time for the Earth to rotate once on its axis (one day), or the completion of one orbit of the Earth around the sun (one year). We can also use the decay of radioactive substances to measure time.

Two of the activities in Module 1 ("Measuring the Period of a Pendulum" and "Building a Drip Timer") require the use of a standard timekeeping device (a watch or clock with a second hand) to calibrate an unconventional timekeeping device. These activities suggest ways of using the units "number of swings of a pendulum" or "number of drops" to measure time.

Timing moving objects using arbitrary units based on the period of a pendulum or drops from a bottle will lead to the same conclusions as observations using standard intervals (seconds and minutes) determined by a watch. Mass-produced modern timepieces are much more convenient to use than a pendulum, but remember—Newton didn't have a stopwatch, because they had not been invented yet!

One should bear in mind that time measures are not absolute. The work on special relativity by Albert Einstein predicted that time changes as velocity changes. These theoretical predictions have been experimentally confirmed by orbiting highly accurate atomic clocks. Atomic clocks have a high degree of accuracy because they are regulated by the natural vibration frequencies of a system of atoms. These frequencies are so stable that some atomic clocks gain or lose no more than a few seconds in 100,000 years. Traveling at about 27,300 kilometers per hour (17,000 miles per hour), orbiting clocks slowed by exactly the amount of time predicted by Einstein's theory.

◆Mass

Detailed discussions of what determines mass, and how mass and weight are different, are included in later readings. A definition of mass which is unique to mechanics relates mass to an object's inertia (or "difficult-to-moveness"). Other ways of defining and measuring mass are considered in Module 2.

◆Accuracy in measurement

All measurements, however carefully they are performed, include some degree of uncertainty or inaccuracy. The precision of the measuring device, as well as the skill of the person using the device, affects accuracy. In many cases, a small amount of inaccuracy in measurement is not a problem. For example, most people have found that their weight on a scale at home disagrees with the weight determined on their doctor's scale. An error of a pound or two when measuring a person's weight usually is not a cause for concern, but if a butcher's scale indicated that a two-pound steak weighed three pounds, the customer would be very concerned!

Similarly, in scientific investigations small errors in measurement may be acceptable in some cases, but a cause for concern in others. Avoiding unnecessary or careless errors while making measurements is important, but some degree of inaccuracy in measurement is inevitable.

Direct and accurate measurements of length, time, and mass are not always possible in scientific investigations. In many cases, an estimate is the best information available. Estimates may be made using indirect means of measurement (such as the rangefinder in a camera) or approximate direct measurements (such as pacing off a distance rather than using a tape measure).

READING 2

Identifying Experimental Variables and Controls

A **variable** is a factor that *may* affect the outcome of an experiment. Identifying factors that may affect the outcome of experiments is an essential skill for all scientific investigations. However, experiments have different possible sets of variables, and students often have difficulty deciding which variables are important for a particular experiment. No experiment is likely to test *all* of the possible variables that may affect the results.

For example, Activity 2, "What Can Change the Period of a Pendulum," identifies two variables (string length and mass of pendulum) for students to test. There are many possible variables that are *not* tested in this activity, including the point of release and the thickness of the string.

The second pendulum activity also illustrates how scientists can design an experiment to discover which variable controls the period of a pendulum. In order to isolate and test the effects of one variable, a scientific experiment must *alter only one variable at a time* and include a **control**.

A **control** is the basis for comparison to all other observations. All the results in the second activity are compared to the observations of a standard pendulum, which serves as the control for the experiment. The standard pendulum has a 50-g weight and is suspended from a 75-cm string. The standard pendulum is always set in motion the same way: by pulling the weight back 10 cm for each trial.

By systematically modifying the pendulum, studying one variable at a time, and comparing the results with the period of the control (the standard pendulum), most experimenters will reach the correct conclusion for Activity 2: the length of the string supporting the pendulum is the variable that determines the period of a simple pendulum.

READING 3

Concepts of Dynamics: Newton's First Law of Motion

◆Historical perspective

The study of the causes of motion or changes in motion is called **dynamics.** We now take for granted the idea that matter behaves in predictable ways when acted upon by a force. But the behavior of matter in motion had not been systematically studied until Galileo Galilei (1564–1642) and Sir Isaac Newton (1642–1727) began their inquiries into the motion of the planets. Their work led to a theory of gravitational attraction and Newton's three laws of motion.

For 200 years after Newton published his *Principia,* these laws of motion were accepted as the best explanations of how matter in motion behaves. In almost all cases examined, bodies in motion behaved exactly as predicted. However, discrepancies were noted between the experimental results obtained and the outcomes predicted by Newton's theories when the laws of motion and gravitation were applied to very massive objects such as stars, or very small objects such as atoms, or objects traveling at velocities approaching the speed of light ($3x10^8$ m/s or about 186,000 miles per second.)

The effort to account for these discrepancies gave rise to two new ideas: the general theory of relativity and quantum mechanics. The general theory of relativity proposed by Albert Einstein (1879–1955) modified and extended the work of Newton to account for the special case of very massive, high-speed objects. Quantum mechanics, the field of physics which describes the behavior of atoms and subatomic particles (those particles of matter smaller than atoms), developed due to the contributions of many scientists.

The "new physics" of relativity and quantum mechanics does not disprove or reduce the importance of Sir Isaac Newton's work. Newton's laws of motion accurately describe all types of motion studied in these activities; scientists and engineers use his principles to calculate the paths of rockets that orbit the Earth or go to other planets. In fact, the results that Newton's laws predict differ from those of relativity and quantum mechanics only in the types of cases mentioned above. In all other cases, Newton's laws are as reliable as the new physics. It is important to remember, however, that Newton's laws are not adequate to describe the behavior of the largest or smallest objects in the universe, or of objects traveling at extremely high speeds.

◆Newton's first law of motion: The law of inertia

Newton's first law of motion can be stated:

> An object at rest tends to stay at rest, and an object in motion tends to stay in motion in a straight line and at a constant speed unless acted upon by an unequal force.

Predicted path of moving ball
(No unequal forces present)

One way to remember this relationship is to think of lumps of matter as being lazy—a mass that is resting will not move unless you give it a shove, and a mass that is coasting along will not alter its course unless you push it or pull on it. A force is required to change the motion of any object. One way to define force is: a force is a push or a pull.

The first part of Newton's law seems to make sense—inanimate objects do not begin moving spontaneously. However, the second part of the law seems to contradict our experience with objects moving near the surface of the Earth. Objects that are moving *do* tend to stop without any obvious action on our part, and usually *do not* move in a straight line.

This seeming disagreement between our experience and the first law can be explained if *all* the forces affecting a moving object are accounted for. What really prevents continuous straight line motion is a *net unequal force* produced by the combination of many forces acting simultaneously. Unless each force acting on an object is canceled out by some combination of the other forces present, the object will be in motion. Identifying *all* of the forces affecting a moving object is difficult, however; many forces are easily overlooked because they are "invisible"—they are so pervasive that we tend to take them for granted and forget to look for them. Such forces can be considered "hidden forces."

To briefly examine how these forces affect motion, imagine the path of a ping-pong ball thrown parallel to the floor. Rather than traveling in a straight line, it curves toward Earth.

Let us try to analyze why the ball curves toward Earth. The arrows in the following diagram represent some of the forces acting on the ball.

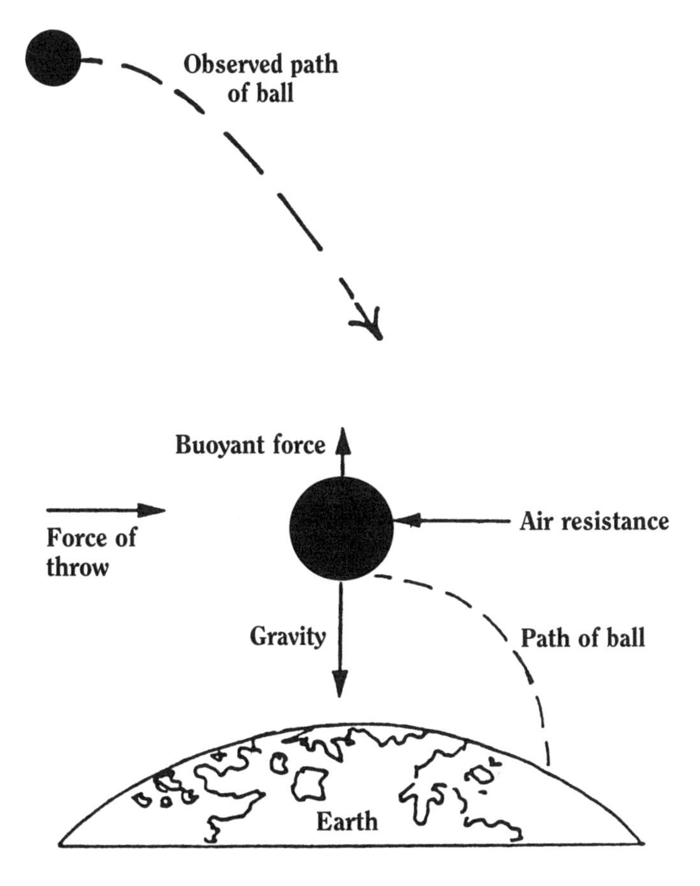

The person throwing the ball gives it the push (force) needed to get the ball moving. This force stops as soon as the ball leaves the person's hand. Friction between the ball and the air (called *drag* or *air resistance*) pushes in the opposite direction when the ball begins to move, causing the ball to slow down.

The gravity of Earth pulls the ball down strongly. The Earth's gravitational force is opposed by the buoyant force of the air. (This buoyant force is the result of air's tendency to exert an upward force on objects in the air. The buoyant force holds up the Goodyear Blimp, but the ping-pong ball's mass-to-volume ratio is too large for it to float on air like the blimp does.)

Newton's great insight was that *in the absence of unequal forces, continuous straight-line motion will occur.* Deviations from straight-line motion near Earth are generally produced by the net effects of friction, gravity, and buoyancy. Other not-so-obvious hidden forces are also acting on the moving ball, but these smaller pushes often cancel each other out because they act simultaneously in different directions. The net effect of these forces is so much smaller than the effect of gravity that they may be disregarded in this example. It is impossible to provide students with perfect examples of uniform straight-line motion as predicted by the first law. Newton's prediction is correct, but the hidden forces combine to produce an unequal net force in all cases. We are not able to reduce friction in mechanical objects to zero, and conducting science classes at zero gravity would require the use of a space ship. Therefore, we must rely on experimental designs that minimize the net effects of multiple forces acting on the object being studied.

The activities in these modules attempt to simplify the analysis of motion by reducing friction as much as possible and by canceling out the net effects of gravity whenever possible. Additional small hidden forces not shown in the ping-pong ball example are always present, but in practical terms they do not affect the results because the measurement techniques used for the activities are not sensitive enough to detect them. However, students should be made aware of the problems of working on Earth when attempting to study uniform motion, and should be encouraged to identify *all* of the forces affecting a moving object, even if these forces cannot be accurately measured.

READING 4

Defining Force

◆Introduction

Each of Newton's three laws of motion describes a different aspect of how forces interact with matter. Knowing how scientists define force is essential for understanding the laws of motion.

Human beings are constantly acted upon by external forces such as gravity. Every move we make exerts force. Many people simply define force as "how hard you push or pull" on an object. This definition is partially correct, but it is not sufficient for scientific use. Scientists studying a force consider not only how hard but the *direction* of the push or pull.

◆Force

A **force** is a push or a pull in a particular direction. When describing forces, both **magnitude** (how much push) and **direction** (which way the push is going) are specified. Forces are invisible; they cannot be directly observed, but the presence of forces can be inferred by observing how objects move or change configuration.

The effects of force can be demonstrated using the spring from a ball point pen. Squeeze the spring between your thumb and finger. The harder you push on the spring (the more force you exert), the more you compress the spring. Squeezing harder increases the *magnitude* of the force. The *direction* of the force you exert on the spring is determined by the direction your fingers move as you squeeze them together.

The following diagram shows several identical springs being compressed by pushes (forces) from above. The forces are represented by arrows called **vectors**. The vector points in the direction of the force. The magnitude (strength) of the force is represented by the length of the vector. A long vector represents a hard push (large force) which is able to compress the spring a great deal.

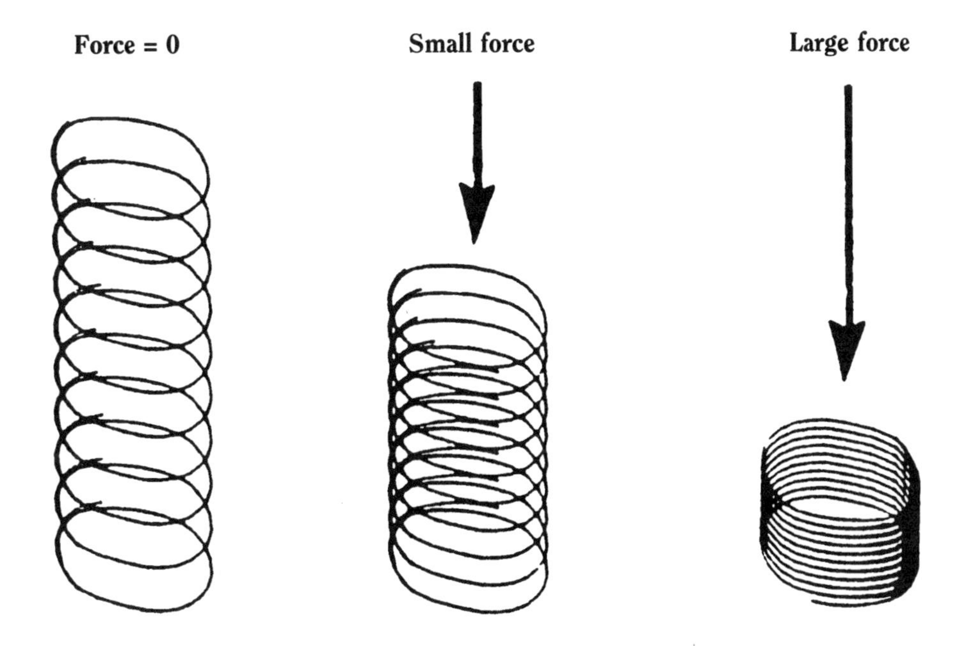

◆Units of force

Any numerical treatment of force requires the use of some standard unit in order to determine how much of the force there is. The standard unit of force is the **newton.**

There is an easy way to remember how much force is equal to one newton. Sir Isaac Newton, for whom the unit is named, is said to have realized the nature of gravity when he saw an apple fall from a tree. A typical apple has a mass of about 100 g. The gravity acting on an apple will cause it to accelerate toward the Earth at about 10 m/s^2. These two facts define the newton:

> One newton is equal to the amount of force required to accelerate a mass of 100 g at 10 m/s^2.

By thinking of Newton's apple falling toward Earth, the meaning of one newton of force (usually written as 1 N) is clear: one newton is equal to the amount of force required to accelerate an object having the mass of a typical apple at 10 m/s^2. The object does not necessarily have to be moving toward Earth, however; acceleration can take place in any direction, depending on the direction of the force acting on the object.

Spring scales are often used to measure force in newtons. Spring scales allow experimenters to determine the magnitude of horizontal forces such as friction, as well as the effects of vertical forces such as gravity. If the spring scales that you use to measure forces are calibrated in grams rather than newtons, you can convert the reading to newtons by dividing the reading in grams by 100. On Earth, a reading of 100 g on a spring scale indicates a force of about 1 N.

READING 5

Using Arrows to Represent Force Vectors

◆Introduction

In Reading 4, arrows were used to represent the forces acting on a spring. A long arrow represented a large force, and a shorter arrow represented a smaller force.

Arrows such as these are called **vectors**. Vectors can be used to represent both the *strength* and *direction* of a force.

◆What is a vector?

A vector is an arrow than can be used to represent a force's strength (magnitude) and direction. Every vector is composed of two parts:

a "head" that shows the *direction* of the force,

and a "tail" that shows the *strength* of the force.

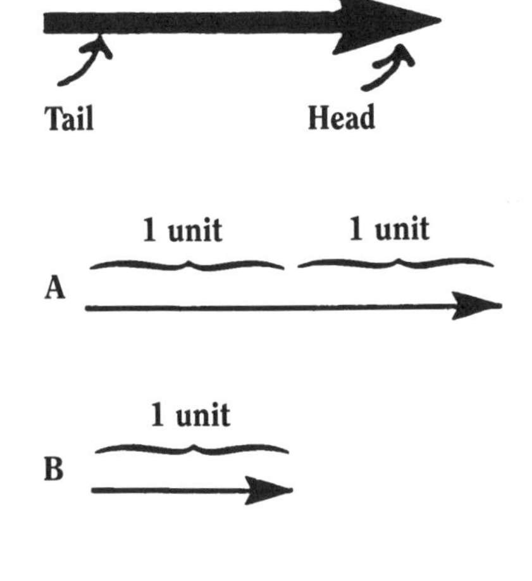

The head of a vector representing a force points in the direction that the force is acting. Vectors A and B both show forces pushing or pulling from left to right.

The longer the tail, the stronger the force. The tail of vector A is twice as long as the tail of vector B, so vector A represents a force that is twice as strong as the force represented by vector B. The units could be newtons, since newtons are used to measure force, but there are other units that are also used to measure force, just as distance can be measured in meters or in feet.

◆Using vectors to represent multiple forces

Objects often have more than one force acting on them at the same time. Determing the net effect of multiple forces can be difficult. However, by drawing an arrow for each force vector acting on an object and combining all the vectors, scientists can predict the **resultant** force (or net force) acting on the object. If the resultant force is greater than zero, the object will tend to move in the direction of that force.

The following diagram shows an example of multiple forces acting on a ball. In the diagram, the ball is resting on the ground. It is acted upon by a downward force of gravity and an upward force of support from the ground. These two forces are equal in magnitude (the tails of the vectors are the same length) and are opposite in direction (the heads of the vectors are pointing 180 degrees apart), so they cancel out and the ball does not move.

Downward force of gravity

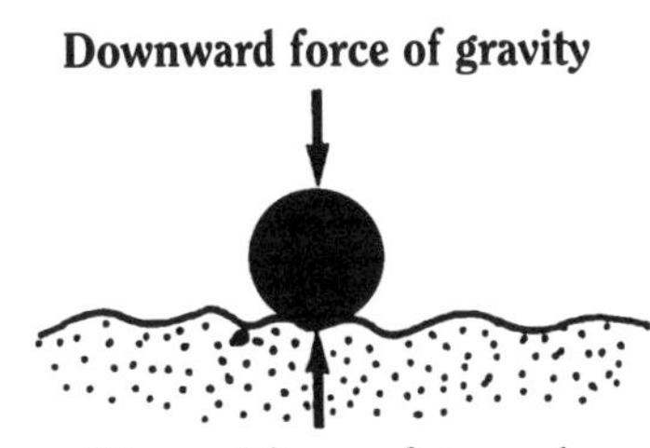

Upward force of ground

In the next diagram, a soccer player kicks the ball, applying a large force (5 units) on the ball from left to right. The force of the kick is opposed by a much smaller force (1 unit) of friction acting in the opposite direction. The upward force of support and the downward force of gravity cancel out, just as in the last example. To find the net force that this produces, you subtract the length of the smaller

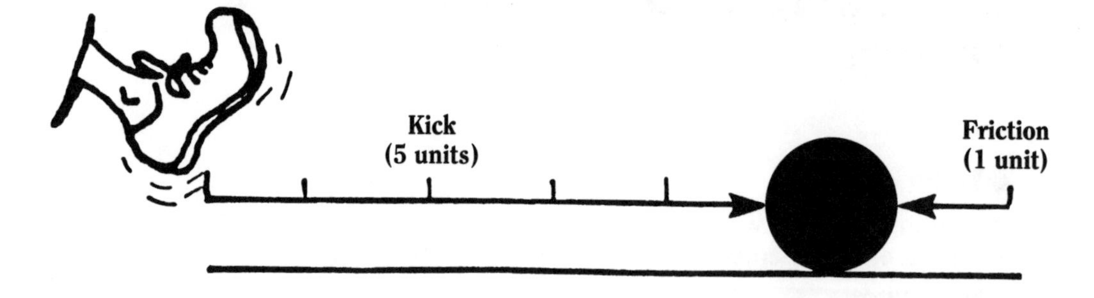

vector (friction) from the length of the larger vector (the kick).

This gives a net force of 4 units magnitude to the right. The ball will move in that direction.

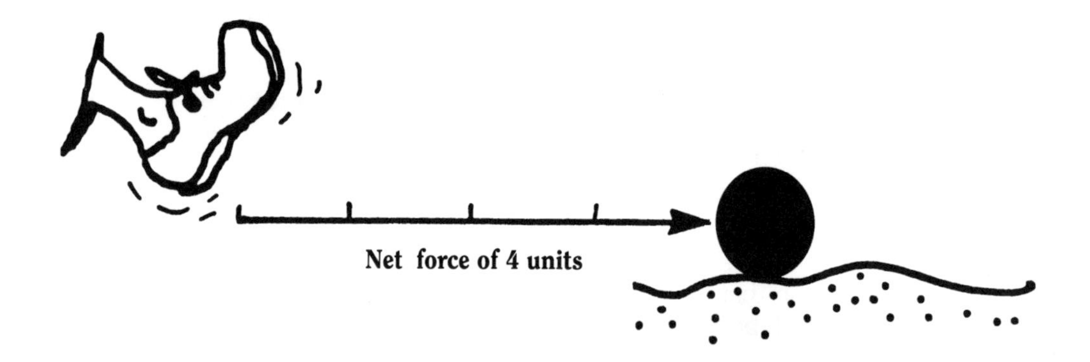

When vectors are used to represent forces that do not meet head-on as in the previous example, you can still add the vectors and find the resultant force.

Consider the case of a ball being pushed off a table. The ball is being acted upon by a force parallel to the surface of the table, and also by the force of gravity, which is perpendicular to the table. How could you predict its motion?

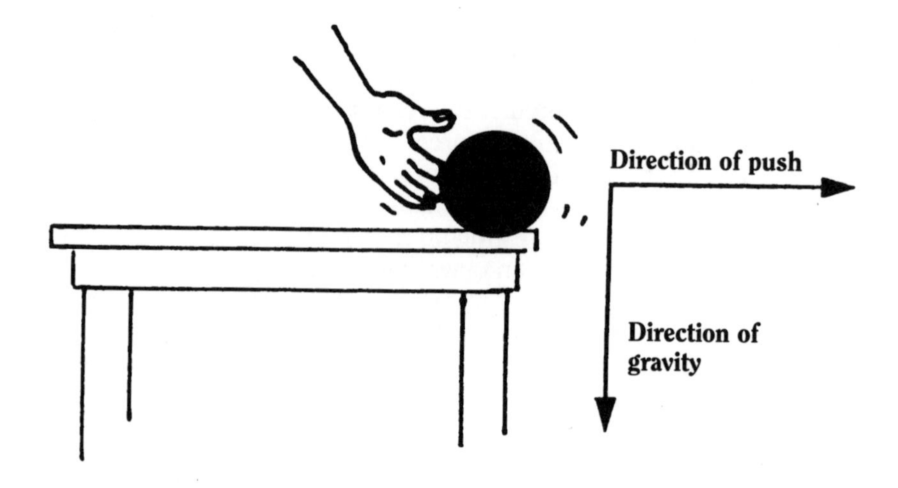

In this case, you add the vectors by rearranging them so that they meet in a "tail to tail" orientation. Then draw a parallelogram using the two vectors as adjacent sides of the figure that you draw.

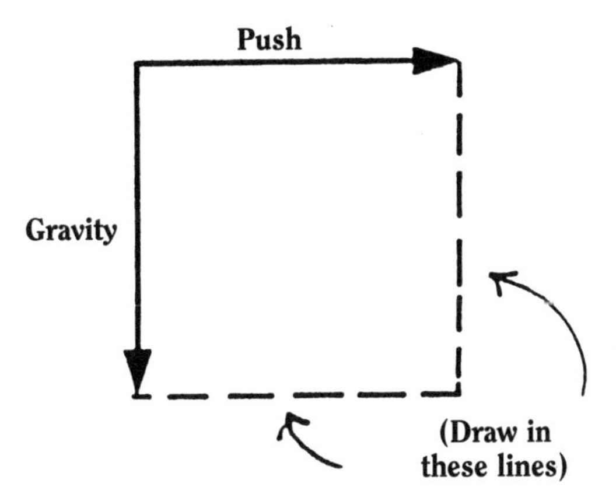

Now draw a diagonal from the point where the two tails meet to the opposite corner. This diagonal shows the resultant force. In this case, the resultant force is downward and from left to right.

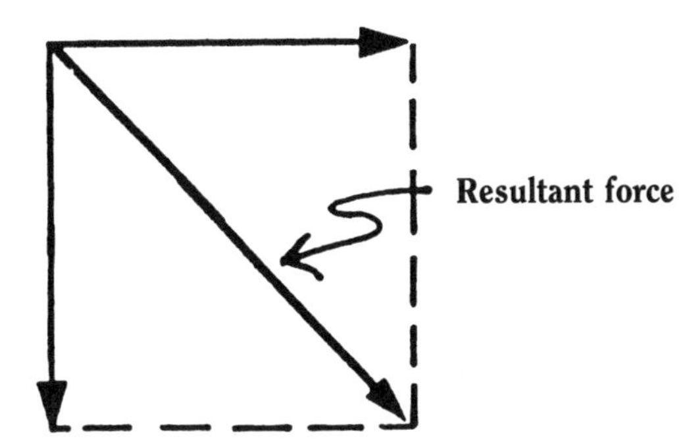

Vectors can represent any quantity that has both magnitude and direction. Vectors are often used to represent velocity and acceleration, as well as force. They are particularly useful in analyzing motion, because a separate vector can be assigned to each force acting on an object. By adding the vectors, the net or resultant force can be obtained. The path of a moving object corresponds to the direction of the resultant. The speed with which it moves corresponds to the magnitude of the resultant. If all of the forces acting on an object can be identified and represented with vectors, its path can be accurately predicted.

READING 6

Three Definitions of Mass

Mass is a fundamental property of matter determined by the quantity of matter present in an object. The total complement of subatomic particles (which includes electrons, protons, neutrons and other particles of matter smaller than atoms) determines how much matter is present in an atom. A chunk of matter large enough to see may consist of hundreds of billions of atoms of many different elements, but the mass of the chunk is still determined by the mass of all the subatomic particles present. The knowledge that matter is composed of electrons, protons, neutrons, and other subatomic particles, each possessing the property "mass," gives the definition:

*The **mass** of an object is the quantity of matter in the object.*

Determining the mass of all the subatomic particles in an object would be one way to determine the mass of an object, but this is impractical. Instead, scientists measure behaviors of matter which are affected by an object's mass in order to determine the amount of mass an object possesses.

The most familiar of these mass-related behaviors is measured using a two-pan balance. By comparing an unknown mass to a set of standardized metric masses on a balance, we can determine the mass of an object in terms of the standard units of grams or kilograms. A second definition of mass, based on this behavior of objects on a balance, can be stated:

*The **mass** of an object is the total mass of standardized mass pieces required to balance the object on an equal arm balance.*

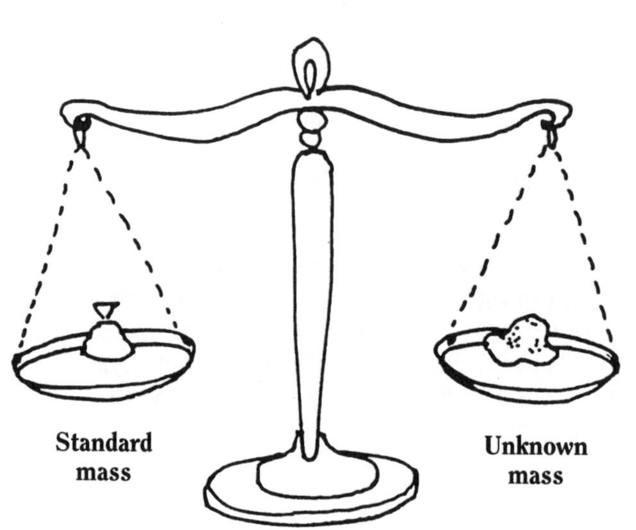

Another mass-related behavior of matter is the property of "difficult-to-moveness," or **inertia**. Objects that possess a large amount of mass (many subatomic particles) are harder to move than objects with a smaller amount of mass. This mass-determined property of "difficult-to-moveness" leads to a third definition of mass:

*The **mass** of an object is the resistance to a change in the motion of the object.*

Determining an object's mass by measuring its "difficult-to-moveness" requires the use of an inertial balance. This is a special type of balance that measures the inertia (or "difficult-to-moveness") of an object whose mass is not known. The mass-measuring device used to determine the relative mass of several objects in Activity 7 is a simple inertial balance.

READING 7

The Magnitude of Motion: Speed

◆Introduction

Everyday descriptions of motion tend to be vague. A speeding car may be described as "really moving" or a good fastball in baseball may be described as "going like a bullet." When quantitative descriptions of motion are used, they are often simply numbers such as "55" read from a speedometer of a car.

A physicist would want to know several facts in order to describe the motion of an object:

- *How far* did the object move while it was being observed?
- *How much time* passed while the object's motion was observed?
- *In what direction* did the object move?
- *What changes* in rate or direction of motion were observed?

◆Computing speed

When we ask about the **magnitude** of something we are asking how much of something there is. **Speed** tells how fast an object travels—how much time is required to cover a given distance.

Determining how far an object moves can be done by comparing its final position to some reference point. Scientists refer to the distance that an object has moved as its **displacement**.

The time interval or "how much time" aspect of motion can be stated in any convenient standardized unit of time. The motion of a jet airplane could accurately be described in terms of how far it travels in a second, how far it travels in 200 swings of a pendulum, or how far it would travel in a year. Hours or minutes are usually the preferred units for describing the motion of a jet, simply because it is much more convenient to say that it takes one hour to fly from New York to Cleveland than 3600 seconds or 1/8760 of a year.

Mathematically combining the "how far" measurement with the "how much time" measurement produces a single number which is a highly useful quantitative descriptor of motion: speed.

Speed is the distance an object moves (its displacement) divided by the time during which the displacement occurred. The distance-time relationship can be stated:

$$\text{Speed} = \frac{\text{distance moved}}{\text{time interval}}$$

$$\text{Speed} = \frac{\text{displacement}}{\text{time interval}}$$

It is important to emphasize that, in strict terms, displacement and distance moved are not identical. *Displacement* is the net distance moved from the starting point. If a man walks 2 m south and then 5 m north, his total displacement is 3 m north. The *distance moved* is the total linear distance, 7 m. Since speed does not take into account direction, the distance moved is the proper quantity to use when calculating speed. Velocity, however, is a vector quantity and is calculated using displacement, since for displacement direction of travel is a factor. For these modules and readings, straight-line motion is used in the examples so that the concepts can be presented without the perhaps confusing distinctions between speed and velocity and displacement and distance moved. Students should, however, be made aware that these distinctions exist and are important.

Speeds can range from zero (for non-moving objects) up to the speed of light ($3x10^8$ m/s, or 186,000 miles per second). Speed indicates the magnitude, but not the direction of motion.

For these activities, the preferred units for stating speeds are **meters per second.** For other purposes, using other metric units such as kilometers per minute or kilometers per hour may be more convenient for describing the speed of an object.

In order to simplify analysis of motion, most laboratory studies of speed are based on observations of objects traveling in a straight line. This allows observers to describe the displacement of a moving object as if it were traveling along the horizontal axis of a graph.

◆A numerical example: Using displacement to calculate speed

The following diagram illustrates how horizontal displacement can be used to determine the speed of an object moving in a straight line.

The diagram shows an air traffic controller in the tower at position 0. She spots an airplane taxiing across the 1000-m runway marker. (Call this

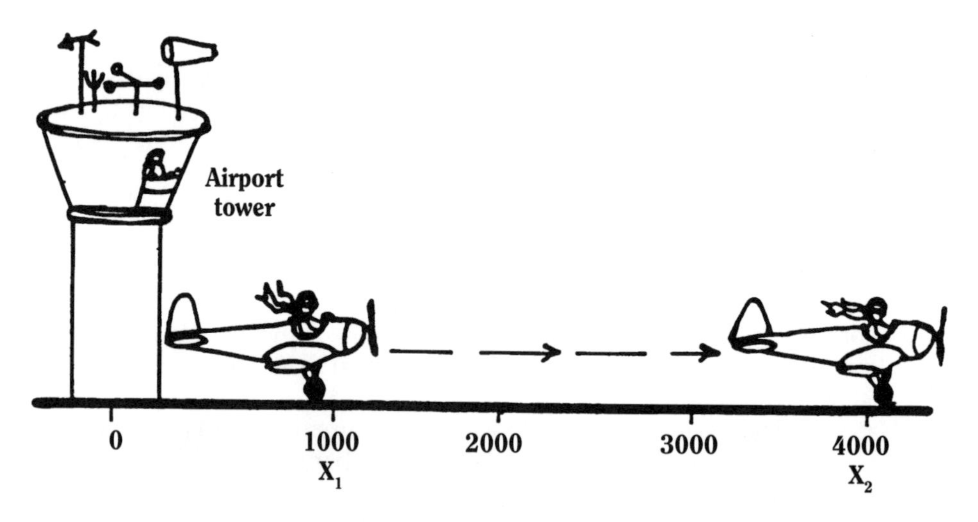

postion x_1.) She observes that after moving for five minutes at a constant speed, it reaches the 4000-m runway marker. (Call this point x_2.)

For these conditions, the motion can be treated as though it occurs on the horizontal axis (x-axis) of a graph, and stated algebraically as:

displacement = final position – starting position

displacement = $x_2 - x_1$

Substituting the numerical values for x_2 and x_1 from the diagram,

displacement = 4000 m – 1000 m = 3000 m.

The equation for speed in a straight line may be stated:

$$\text{Speed} = \frac{\text{final position} - \text{starting positon}}{\text{time interval}}$$

$$= \frac{\text{displacement}}{\text{time interval}}$$

For the airplane shown in the diagram, the displacement during a five-minute time interval was 3000 m. By restating the time interval in terms of seconds (5 minutes = 300 seconds), the speed can be computed as follows:

speed of the airplane = 3000m / 300 s
= 10 m/s

For many scientific purposes, meters per second are the preferred units for describing speed. However, the speed can also be stated in the following equivalent units:

speed of the airplane = 3 km / 5 minutes
= 0.6 km/min.

speed of the airplane = 3 km / 0.083 hours
= 36 km/hr

◆Using delta notation for intervals

Describing motion frequently involves calculating different types of intervals. The calculation of an object's speed in the previous section required the determination of displacement (the final position minus the starting postion, or $x_2 - x_1$) as well as the time interval (the seconds during which the motion occurred). The Greek letter delta, written as Δ, is often used as an abbreviation for the phrase "calculate an interval of ___." This notation can be applied to many cases. For example, $x_2 - x_1$ can be abbreviated as Δx. An interval of time can be abbreviated Δt. Changing velocity can be shown as Δv.

Using the delta notation may seem to increase the abstraction of certain formulas. However, we use notation such as a/b to represent "divide a by b" or $\sqrt{t}$ to represent "take the square root of t" without giving it a second thought. Using the notation quickly becomes as automatic as using the other operational symbols, and allows many equations to be stated in briefer form. For example, speed can be represented:

$$\text{speed} = \frac{x_2 - x_1}{t_2 - t_1} = \frac{\Delta x}{\Delta t}$$

The delta notation is read, "speed is equal to delta x divided by delta t" or "speed is equal to the change in x (position) divided by the change in t (time)" or "speed is equal to the interval of x divided by the interval of time." This notation is used throughout the modules on mechanics.

◆Speed versus distance: What does "constant speed" mean?

Formal definitions of speed based on time interval and displacement along an x-axis are useful for performing mathematical analyses of moving objects. A more intuitive approach to defining the speed of an object requires thinking in terms of "How far did it go?" and "How much time did it take to get there?" Using this approach, **constant speed** can be operationally defined as follows: an object traveling at a constant speed will move the same distance during every equal time interval.

A good way to visualize what is meant by this definition of constant speed is to imagine that you are riding in a moving car and dropping a spoonful of pancake batter out the window every second. The position of the spot of batter on the road would show approximately where the car was located at a particular moment.

The distance between the spots of batter on the road would show how

Batter dropped at one-second intervals

5 m 5 m 5 m 5 m

Constant speed of 5m/s

far the car traveled during 1 s. If the car is moving at a constant rate of 5 m/s, a person walking along the road would find spots of batter spaced about 5 m apart along the road.

As long as the car continues moving at a constant speed, the drops of batter will land on the road the same distance apart. (A similar situation is illustrated in the videotape in Activity 15 for a train traveling at a constant speed; the effects of changing speeds are also illustrated.)

If the car were moving at 10 m/s, dropping the batter each second would leave a trail of spots exactly 10 m apart. For a car moving at a constant speed, when the batter is dropped out the window at equal time intervals, the distance between the drops of batter is determined by the speed of the car. These observations support the operational definition for constant speed: an object traveling at a constant speed will move the same distance during every equal time interval.

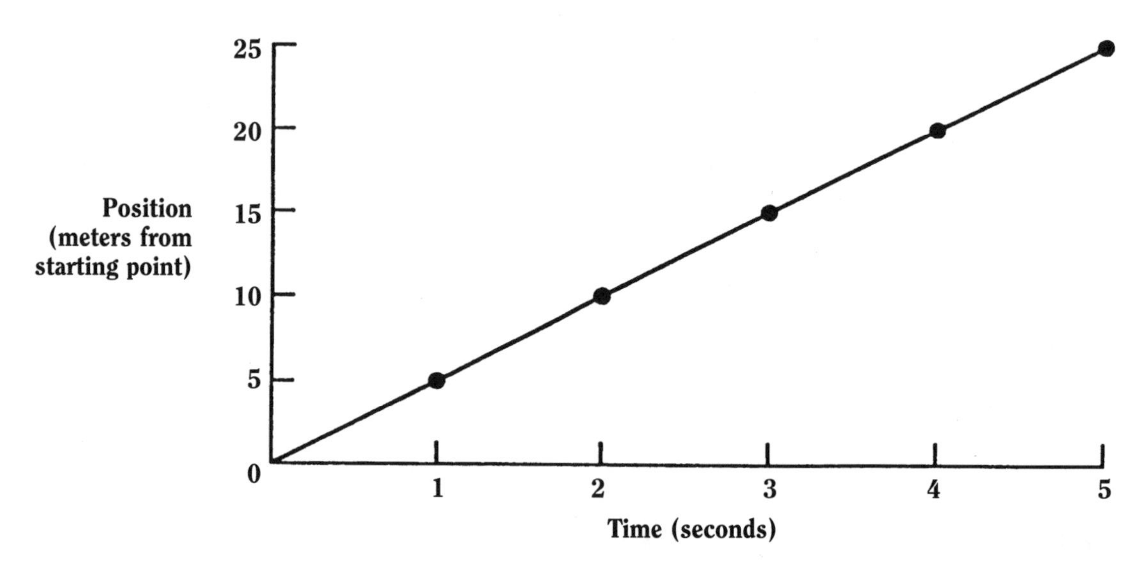

A more formal way to show this relationship employs a position-time graph. The positions of the car moving at a constant 5 m/s can be shown on a graph, above. Take note that in this simplified graph, the car is timed after it has achieved its full speed, 5 m/s; it does not build up from 0 m/s.

A graph such as this allows one to predict the position of the car at any given time.

◆Speed versus velocity: How are they different?

Portable electric radar guns are often used by television crews to measure the speed of the ball during tennis matches and baseball games. The announcers who receive this information use the terms *speed* and *velocity* interchangeably to describe how fast a player is serving or throwing the ball. Physicists watching the game know that velocity is *never* the same as speed. Speed is a measure of the magnitude of motion, but velocity is a measure of both the magnitude and the direction of motion. In other words, speed only tells us how fast, but velocity tells us how fast and in what direction. Notice that if we know the velocity of an object, we automatically know its speed, since speed is one of the two things velocity tells us. However, if we only know an object's speed, we can't say what its velocity is until we determine in what direction it is moving.

The formal scientific distinction between speed and velocity is important for describing certain types of motion such as projectile motion. In these modules, all activities requiring speed determinations use objects moving in a straight line in order to simplify data analysis. In these activities, the direction in which the object is moving does not affect the results, so the term *speed* is used in all cases for describing objects' motion.

READING 8

An Intuitive Approach to Defining Acceleration

◆Introduction

Research shows that students who study Newton's second law of motion (Force = mass X acceleration) can usually reproduce the formula correctly on tests and use it to compute numerically correct answers to problems. However, many of these students cannot predict simple motions of actual objects because they are not able to relate the accelerations they experience every day to the terms in the equation.

Module 3 uses hands-on activities to clarify students' intuitive knowledge of acceleration. Scientifically useful operational definitions of acceleration are based on the results of the activities. These operational definitions are used along with numerical definitions of acceleration to demonstrate how Newton's second law predicts the behavior of moving objects in the real world.

◆Relating acceleration to students' experiences

Each of us has an intuitive understanding of acceleration based on our everyday experiences. We undergo acceleration in automobiles each time the driver steps on the gas pedal or applies the brakes. This type of motion is so familiar that we automatically lean forward or backward, or brace ourselves in our seats to counteract the forces that produce accelerations.

Most people think about acceleration in terms of the sensations they experience while speeding up or slowing down. Translating and refining students' practical awareness of acceleration in a way that remains intuitively reasonable and becomes scientifically useful should be a major goal for physical science teachers.

◆Acceleration: What is it? What causes it?

All matter in the universe is doing one of three things: (1) it is sitting still, (2) it is moving at a constant speed, or (3) it is changing speed and/or direction of motion. Acceleration is the term we use to describe the third case. A simplified definition of acceleration will be used in this module: Acceleration is the rate of speeding up or slowing down.

Some texts introduce the concept of changes in motion by differentiating between acceleration (speeding up) and deceleration (slowing down), positive acceleration (speeding up) and negative acceleration (slowing down), and by defining centripetal acceleration (moving along a curved path.) Using all of these terms immediately is not essential if the examples of motion being studied are carefully selected. Introducing these distinctions when beginning the study of acceleration may confuse students. However, it is important to realize that the acceleration term, *a*, in F = ma applies to objects that are slowing down as well as those that are speeding up.

In order to simplify the study of acceleration, this module mainly uses examples in which an object is moving in a straight line and speeding up. Under these conditions, the direction of motion need not be accounted for.

Acceleration can only occur when a force acts on an object. The force must be large enough to overcome the object's *inertia* (or "difficult-to-moveness") and any *frictional forces* present. A gnat flying into a bowling ball sitting on the floor cannot cause it to move. The force the gnat can exert on the ball, though unbalanced, is too small to overcome the combination of the ball's inertia and the friction between the ball and the floor.

A human bowler's hand and arm can easily apply enough force to change the speed of the ball from zero to a final speed of several meters per second. All acceleration takes place before the ball is released; the push (force) exerted by the bowler can only act while the ball is touching the hand. After the ball is released it travels at a fairly constant speed until it strikes the pins, as predicted by the first law of motion.

Sometimes it is easy to *infer* that acceleration is occurring. A drag racing car accelerates from a standstill to over 200 miles per hour and covers a quarter-mile course in less than seven seconds. The car's engine provides the unequal force that produces the acceleration. The change of speed is obvious in this case.

There are many situations in which it is difficult to judge whether or not acceleration is occurring. Anyone who can see an entire drag strip can tell whether or not a car is speeding up over the quarter-mile distance. But if you were looking through a narrow window that only allowed you to see the car move one car length down the track, could you tell for sure if it was speeding up or slowing down? Probably not; naked eye observations are not always reliable if an action occurs in a short period of time, or if the motion is viewed in a limited way (such as looking through some type of narrow window). Therefore, the first problem one must confront when studying acceleration is to answer the question, "Is the object speeding up, slowing down, or maintaining its speed?"

◆Operational definitions

Operational definitions offer one alternative to the traditional instructional approach of defining acceleration mathematically. An operational definition has two parts. It tells you:

1. What you do or what *operation* you perform.
2. What you *observe*.

An example of an operational definition that a chemist might give for oxygen is:

> When you place a glowing splint into a container of gas (*what you do*), the splint will burst into flame (*what you observe*) if oxygen is present.

This operational definition does not tell you everything about oxygen. It does not tell you about the number of protons and neutrons found in oxygen, or that oxygen combines with hydrogen to form water. The operational definition does, however, provide one way to answer the question: "Is oxygen present in a container of gas?"

Similarly, using an operational definition approach to studying acceleration allows students to decide whether or not acceleration is occurring by observing objects in familiar contexts, and by analyzing how they are moving. Not all of the definitions of acceleration given in this module are operational definitions, but several activities in this module are designed to illustrate and reinforce the concepts leading to operational definitions of acceleration. These activities parallel the approach taken in previous activities for investigating constant speed.

◆Verifying that a change of speed occurred: Three operational definitions of acceleration

Deciding whether or not acceleration is taking place sometimes requires the use of photographic techniques that freeze the motion for later analysis. However, studying acceleration is possible without using elaborate equipment. The question "Is the object speeding up or slowing down?" can sometimes be answered by using simple timers or accelerometers. Observations using these devices allow the following statements to serve as operational definitions of acceleration:

1. Acceleration is occurring if an accelerometer's indicator moves away from the rest position.
2. Acceleration is occurring if an object travels different distances during a series of equal time intervals.
3. Acceleration is occurring if there is a change in the time interval required to move a constant distance.

These are not complete definitions of acceleration, but they do provide concrete, non-mathematical methods of determining whether or not acceleration is occurring on the basis of direct observations of moving objects.

Why are these definitions classified as operational definitions? Remember that operational definitions have two parts: they give you a task to carry out (an operation) and they tell you the expected result of the observation that you make. Examine the first definition: acceleration is occurring if an accelerometer's indicator moves away from the rest position. For this definition:

"What you do" is operate a simple piece of equipment called an accelerometer;

"What you observe" is the position of the indicator on the accelerometer.

If the indicator moves away from the rest position, acceleration is occurring; if it stays still, no acceleration is occurring.

For the second definition, "what you do" is measure the distance traveled by a moving object at equal time intervals; "what you observe" is whether or not the distance traveled during each equal interval changes. Can you explain "what you do" and "what you observe" for the third definition?

READING 9

Algebraic Representation of Acceleration

◆Introduction

While some students may be comfortable with algebraic representations of acceleration, the formal scientific units of acceleration, meters per second per second, have little meaning in the real world for most students. The commonly used unit for acceleration, m/s^2, is even more abstract and less intuitively accessible; what, students may ask, is a "square second?"

Teachers assigning practice problems must realize that mathematically solving acceleration problems does not prove that a student can relate the intuitive notion of acceleration (speeding up or slowing down) to how an actual physical object behaves while a force is acting on it. Relating hands-on laboratory experience measuring acceleration to the numerical representations of acceleration is essential to building an understanding of acceleration.

◆Units and notation

A numerical value for acceleration may be obtained by subtracting the initial speed from the final speed, and dividing by the amount of time required for the change of speed to occur.

$$\text{Acceleration} = \frac{\text{final speed} - \text{initial speed}}{\text{time elapsed between measurements}}$$

or,

$$\text{Acceleration} = \frac{\Delta\text{speed}}{\Delta\text{time}}$$

All that this number representing acceleration indicates is whether or not the object is changing speed. If the speed of the object does not change during the time interval, the final speed and initial speed are identical. Subtracting identical speeds from one another gives a zero in the numerator, so the value of acceleration for an object moving at a constant speed is always zero, even if the object were moving at the speed of light.

◆Acceleration as related to velocity

In most science text books, acceleration is expressed as the rate of change of velocity. Velocity specifies both the magnitude and direction of motion. Defining acceleration in terms of changing rates of speed is simpler and is also acceptable, as long as the direction of motion does not change. The units for speed are the same as those for velocity, meters per second. For the activities in this module, using velocity to describe motion is not essential, since only straight-line motion is studied. The *Eureka!* videotapes also avoid using the term *velocity*. However, the following derivation is stated in terms of velocity rather than speed, in order to conform with commonly encountered notation.

In standard notation, when the first velocity measurement is done at time t_1 and the final measurement is done at time t_2:

$$\text{Acceleration} = \frac{v_2 - v_1}{t_2 - t_1} = \frac{\Delta v}{\Delta t}$$

where v_1 = initial velocity (meters per second)
v_2 = final velocity (meters per second)
$t_2 - t_1$ = time elapsed between measurements of v_2 and v_1 (seconds)

Therefore, the *units* for acceleration can be stated:

$$\text{Acceleration} = \frac{\text{(meters/second)} - \text{(meters/second)}}{\text{(seconds)} - \text{(seconds)}}$$

Which can be simplified to:

$$\text{Acceleration} = \frac{\text{meters/second}}{\text{second}} = \frac{\text{meters}}{\text{second}^2}$$

Applying this formula to a specific example, suppose that an automobile is observed for a total of 8 s. After 5 s its velocity on a straight, flat road is 25 m/s. After 6 s, its velocity is 30 m/s. What is its acceleration for that time interval?

For this case,
v_1 = initial velocity
= 25 m/s
v_2 = final velocity
= 30 m/s
t_1 = 5 s
t_2 = 6 s

Substituting in the formula,

$$\text{Acceleration} = \frac{v_2 - v_1}{t_2 - t_1} = \frac{30\text{ m/s} - 25\text{ m/s}}{6\text{ s} - 5\text{ s}}$$

$$= \frac{5\text{ m/s}}{1\text{ s}} = \frac{5\text{ m}}{\text{s}^2} \quad \text{or } 5\text{ m/s}^2$$

The numerical value for acceleration in this case, 5 m/s^2, indicates that between the fifth and sixth seconds during which the motion was being observed, the rate of increase of speed was 5 meters per second per second.

This value does not tell us what occurred in the intervals before or after the observations were made. It represents the narrow window through which we view one segment of the car's motion.

◆Formulas relating acceleration to velocity and distance moved

The rate of acceleration is useful scientifically, but is rarely very important in everyday life. We are more likely to want to know the speed of an object or the distance it traveled during a time interval than its rate of acceleration. For example, suppose the rate of acceleration of a toy car rolling down a ramp is found to be 0.5 m/s^2. To comprehend what that rate of acceleration means, we would probably try to relate the rate of acceleration to common experiences by asking how far the car would travel in a minute, or how fast it would be moving after an hour if it continues accelerating at 0.5 m/s^2.

If the rate of acceleration is known, the speed and distance traveled can be calculated for any time interval. The following formulas describe the velocity and distance traveled for an object accelerating uniformly in a straight line:

velocity = acceleration X time ($v = at$)

distance = (1/2) X acceleration X time^2 ($d = (1/2)at^2$)

Now let us examine the motion of our toy car accelerating down a ramp at a rate of 0.5 m/s^2. This seems like a rather unimpressive rate of acceleration. But is our intuition correct? We can now calculate the velocity that the car would achieve if it somehow continued to accelerate at that rate for an hour, and see how far it would travel.

Applying these formulas to that hypothetical situation produces some surprising results. A continuous, uniform rate of acceleration can quickly produce very high speeds and result in large distances being traveled. The following table showing the acceleration of a car is based on the following assumptions:

- The acceleration occurs at a constant rate of 0.5 m/s^2.
- The car is moving in a straight line and is starting from rest.

Acceleration of a toy car

Elapsed time t	Velocity achieved $v=at$	Total distance $d=1/2at^2$	Comparison
1 s	0.5 m/s	0.25 m	A sheet of notebook paper is about 0.25 m long.
10 s	5.0 m/s	25 m	Olympic sprinters cover 100 m in about 10 seconds.
60 s	30 m/s	900 m	This speed equals about 67 mph; the distance is about equal to 9 football fields.
3600 s (1 hour)	1800 m/s	3,240,000 m	This distance is about 2014 miles. The distance from New York to Los Angeles is about 2800 miles.

After only 1 hour at a constant acceleration of 0.5 m/s^2, the car would be traveling 1.12 miles per second (about 4027 mph, or about six times the speed of sound). It would have traveled 2014 miles. If the car continued accelerating at this rate, it would reach the speed of light, about 186,000 miles per second (or 3×10^8 m/s), in less than twenty years. At that speed, in one second it would be able to travel a distance equal to seven trips around the Earth at the equator.

The calculations summarized in the table are numerically correct, but no one is likely to believe that a toy car rolling down a ramp could reach a velocity of 30 m/s. Exceeding the speed of sound with a toy is clearly preposterous. Our experience and intuition tell us that something prevents continuous acceleration from occurring. That something consists of multiple "hidden forces," forces that are easy to overlook when people try to explain motion, such as air resistance and friction between wheels and axles. Hidden forces will be investigated in more detail in a later module.

READING 10

Momentum

◆Introduction

The term **momentum** is commonly used in news articles describing political campaigns or in television commentaries analyzing the prospects of teams entering playoff series. To non-scientists, momentum has come to mean forward thrust. This is comparable to the intuitive idea that most people have of the physical phenomenon of momentum; they might define momentum as the quantity of "bashing power" that a moving object possesses.

Persons using this definition would (correctly) say that a loaded concrete mixing truck has more momentum than a compact car moving at the same speed. They would also be able to predict in a general way the outcome of a collision between a high-momentum truck and a low-momentum car. This informal bashing power definition is similar to physicists' definition of momentum.

◆Defining momentum scientifically

Momentum, to the physicist, is a formal mathematical concept. The momentum of an object is calculated by multiplying its mass by its velocity. This can be stated as an equation:

momentum = mass X velocity

The letter p is commonly used to symbolize momentum, and the equation for momentum can be written:

$$p = mv$$

Momentum provides highly useful information for predicting interactions of objects having different masses and velocities. Knowledge of momentum allows physicists to make accurate predictions of outcomes, rather than just assuming that a large object will bash a smaller object out of the way. In fact, the opposite may be true if the small object is moving fast enough. For example, colliding with dust particles moving at extremely high speeds (an estimated 70,000 m/s or about 155,000 miles per hour) disabled the Giotto satellite photographing Halley's comet.

The third *Eureka!* videotape, *Speed*, illustrates how speed and mass affect the bashing power of balls ranging from tennis balls to cannonballs. Although the term *momentum* is not used in the program, the cartoon characters demonstrate the relationship $p = mv$ in several different contexts during this segment. They show that while it is possible to catch a massive, but slow-moving cannonball, a high-velocity tennis ball could literally knock you over. Both mass and velocity must be considered when predicting an object's momentum.

READING 11

Newton's Second Law of Motion

◆Introduction

Anyone throwing a ball is gaining practical knowledge of accelerated motions. Relating these everyday experiences with objects that are changing speed to the mathematical relationships stated in Newton's second law of motion is the goal of this section.

The second law of motion can be represented in two different ways: in terms of force, commonly stated as F = ma, or in terms of an object's change of momentum. Newton originally stated the law in terms of momentum, which he called the "quantity of motion."

◆Newton's second law of motion

Sir Isaac Newton's second law of motion states:

Force = mass X acceleration.

This equation, commonly written as F = ma, provides a quantitative means of distinguishing between a "big push" and a "little push." It also allows us to predict how different masses will move when acted upon by a certain quantity of force.

Recall the definition of the newton (N), the standard unit of force: a force of 1 newton will accelerate a 100-gram mass (such as a typical apple) at 10 meters per second per second. Substituting this information in the second law of motion, F=ma, gives the equation:

1 newton = 100 grams X 10 meters per second per second.

This word equation means the same thing as the equations:

$1 \text{ N} = 1000 \text{ g m/s}^2$ *or* $1 \text{ N} = 1 \text{ kg m/s}^2$.

What would happen if a force of 1 newton acted on a 200-gram mass (a very large apple) rather than a 100-gram mass? The mass is now twice what it was before. What acceleration will take place?

The second law of motion allows us to predict the acceleration of the 200-gram mass being acted upon by a 1 newton force as follows:

We know that $F = 1 \text{ N} = 1000 \text{ g m/s}^2$ and $m = 200\text{g}$.

Substituting these values in the second law, F = ma, gives:

$$1000 \text{ g m/s}^2 = 200 \text{ g} \times a$$

$$\frac{1000 \text{ g m/s}^2}{200 \text{ g}} = a$$

$$a = 5 \text{ m/s}^2$$

Therefore, we predict that the 200-gram mass will accelerate at a rate of 5 meters per second per second. When the force acting on a mass remains the same, doubling the mass will halve the rate of acceleration.

This calculation shows how the second law of motion allows us to predict the acceleration of an object given the force applied to it. Another important aspect of a force acting on an object is the *duration* of the action of the force. The net effect of a force is determined not only by its magnitude and direction, but also by the amount of time that it acts on an object. A small force acting over a long period of time may eventually cause a mass to move at a greater speed than a very large force that acts for only a brief amount of time.

READING 12

The Third Law of Motion

◆Introduction

Sir Isaac Newton's third law of motion predicts how objects will interact. The law may be stated either in terms of *action and reaction* or in terms of *opposing forces*:

For every force there is an equal and opposite force.

To every action there is an equal and opposite reaction.

If the objects being studied are moving, the "action-reaction" statement may be easier to use; the "opposing forces" statement may be more helpful for examining a static situation, such as a mass sitting on a surface.

◆Equal and opposite forces

For objects that are not moving, identifying equal and opposite forces can be tricky. Though it is easy to imagine that a 100-g object exerts a downward force of 1 N on a desk, the idea that the desk is pushing back in the opposite direction with a force exactly equal to 1 N is very subtle.

You can measure the *downward* force that an apple exerts by suspending it from a spring scale. But how do you measure the *upward* force exerted by the desk? How can we be sure that the desk does not push back with a force of 200 N?

Engineers design bridges and other structures so that they will be strong enough to resist large downward forces without danger of collapse. Some of the questions that engineers must ask when designing a structure include:

• What is the maximum load that could possibly be placed on the structure in normal use? (For a bridge, this might be a bumper-to-bumper traffic jam of heavily-loaded trucks, completely covering the bridge in both directions.)

• What unusual loads might affect the structure from time to time? (A layer of ice on a bridge that is being buffeted by high winds puts abnormal amounts of force on parts of the structure.)

• How strong is the building material? (Bridges are constructed of steel and concrete. Engineers must know how much force a given amount of these materials can withstand without breaking.)

• Does the building material weaken with age? (Iron and steel bridges rust, and the concrete bridge surface can be broken by the weight of trucks and the action of freezing and thawing water.)

Once engineers answer these and other questions, they can design a structure that will be safe to use for many years. Unlike the "Fettucini Physics" structure, it will not suddenly fail. Well-designed structures are capable of exerting more than enough upward force to balance any downward force that is ever likely to be exerted upon them.

◆Action-reaction pairs

Thinking of objects in motion as action-reaction pairs is the easiest way to analyze certain examples of motion. One such case is the firing of a gun. The gunpowder exploding inside a round of ammunition produces a force that acts on both the bullet and the gun. The kick of the gun and the flight of the bullet are an action-reaction pair of motions.

The "Kodak Cannons" activity in Module 5 is designed to demonstrate this type of action-reaction pair. The "cannon" and "bullet" (the film containers) are, for all practical purposes, identical. The force (provided by the carbon dioxide produced by the Alka-Seltzer) is transmitted equally to both the cannon and the bullet. Since the same force is acting on both the bullet and the cannon, and their masses are equal, we can use the third law of motion to predict their motions:

The *action* (the motion of the bullet) will have the *same magnitude* as the *reaction* (the motion of the cannon.)

The action and reaction will occur in *opposite directions* and act on different objects.

The observations of the empty bullet and cannon support these predictions. The bullet and cannon both move about the same distance; therefore we can infer that the action and reaction are of the same magnitude. The bullet and cannon move in opposite directions as predicted.

◆More about action-reaction pairs

A question remains about what happens when a gun is fired: if the force acting on a bullet is equal to the force acting on a gun, how can a bullet travel almost 2000 m at over 500 m/s, while a gun being fired moves very little?

The results obtained by changing the mass of the bullet used in the "Kodak Cannons" activity suggests an answer to this question. If one of the members of an action-reaction pair is much more massive than the other, the distances that the two objects move will be different. The more massive object will not move as much as the less massive object of the pair. The second law of motion, F = ma, can be used to show mathematically why this is the case.

The force acting on the bullet and the cannon can be stated as follows:

$$\text{Force} = \text{mass}_{\text{bullet}} \times \text{acceleration}_{\text{bullet}}$$

and

$$\text{Force} = \text{mass}_{\text{cannon}} \times \text{acceleration}_{\text{cannon}}$$

For this activity, the *same force* acts on both the bullet and the cannon. Therefore,

$$\text{mass}_{\text{bullet}} \times \text{acceleration}_{\text{bullet}} = \text{mass}_{\text{cannon}} \times \text{acceleration}_{\text{cannon}}$$

If the mass of the bullet and the cannon are equal, they will accelerate at the same rate; if the masses are different, the more massive member of the pair (in this activity, the bullet containing water or lead) will accelerate a smaller amount.

Although acceleration is not measured directly for this activity, the distances moved are much smaller for the more massive bullet than for the cannon; this correlates with the predicted motions.

READING 13

Can Objects Break the Laws of Motion?

◆Introduction

Studying the behavior of matter in motion is complicated by "hidden forces" that change the direction and/or speed of moving objects in ways that may surprise an unwary observer. However, if *all* the forces acting on an object are accounted for, the object will be found "not guilty" of "breaking the laws of motion."

◆Hidden forces affecting motion

Newton's first law of motion states:

> An object at rest tends to stay at rest, and an object in motion tends to stay in motion in a straight line and at a constant speed *unless acted upon by an unbalanced force*.

On Earth, the pervasive unseen forces of gravity and friction act on all moving objects. As a result of these unequal or unbalanced forces, the motion of most objects is non-linear and variable in speed, as Newton predicted. Having to account for multiple, unseen, unequal forces acting on an object greatly complicates the analysis of motion.

To reduce the cumulative effects of these unseen forces, most laboratory demonstrations of motion use carefully selected objects moving at low speeds. Objects such as marbles rolling on a track are often used because they roll with little friction, possess a fairly large mass, and have a small volume. Because these types of objects are less noticeably affected by the hidden forces which are always present, they produce motions which are easier for most people to understand.

◆Hidden effects of air resistance

The force of air resistance (also called drag) acting on an object is proportional to the surface area of the object multiplied by the speed squared (in other words, drag $\approx$ surface area X speed^2.) Doubling the speed of a moving object increases the force of air resistance by a factor of four.

There are important practical applications of the relationship between drag and speed. During an oil embargo several years ago, the national speed limit was lowered to 55 mph in order to conserve fuel supplies. Driving at 70 mph uses significantly more gasoline than driving 55 mph, because increasing the speed of the vehicle by 27 percent (from 55 to 70 mph) increases the air resistance (drag) by 62 percent.

The effects of air resistance can be eliminated for all objects by studying their motion in a vacuum. However, producing and maintaining a high vacuum is so technically difficult that vacuum apparatus is rarely used in classrooms. By using smooth, symmetrical objects of low surface area traveling at low speeds, the magnitude of the effects of air resistance can be reduced to levels that are, in practical terms, insignificant.

◆Hidden effects of gravity

Gravitational force cannot be avoided. Even "weightless" orbiting spacecraft are kept in orbit by gravity. If there were no gravity, all satellites would fly away from Earth on a straight line, rather than curving around the Earth. However, since gravity exerts a constant force on an object at any position along a level surface, it is possible to cancel out its effects to a certain extent.

If an object being studied is moving horizontally on a level, smooth, low-friction surface, the constant gravity may have very little obvious effect on its motion along the surface. The object is prevented from falling down by the upward force exerted by the surface, so it travels in a straight line. Minimizing the effects of gravity makes analyzing the motion much easier.

◆How hidden forces interact

The following example using "Newton's Apple" shows how the hidden forces of gravity and air resistance sometimes change an object's motion by working against each other.

Imagine an apple hanging from a tree. It is not moving, so it has a speed of zero, and there is no upward force of air resistance acting on it. Gravity is pulling it toward the Earth with a constant force, and the stem of the apple is pulling it up with a force equal to the weight of the apple.

Force of stem
Speed = 0
Air Resistance = 0
Force of gravity

If the stem of the apple breaks, gravitational force accelerates the apple toward Earth at a rate of about 10 m/s^2. With each passing second, the apple falls faster and faster. As the speed of a falling object increases, the frictional force of the air becomes larger and larger.

Force of stem = 0
Speed increases
Air resistance increases
Force of gravity

If the apple falls far enough, its speed will increase until the upward frictional force of the air exactly equals the downward force of gravity. (Remember, doubling the speed of a moving object increases the force of air resistance by a factor of four.) When these forces are equal, the object continues to travel at that speed until it strikes the Earth.

The speed at which the force of air resistance equals the force of gravity is called the **terminal speed**. Once a falling object has reached its terminal speed, it will continue to travel in a straight line at a constant speed unless it is acted upon by some other force—for example, until it hits the ground.

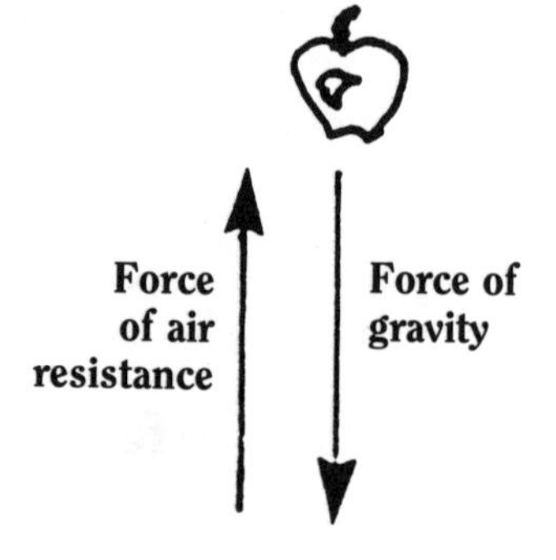

◆Newton's apple: More about how it fell

The apple was chosen for the previous example because Newton reported that he had experienced a flash of insight about gravitational acceleration after observing an apple falling. However, the scale of the drawings for the example is slightly misleading. In order for a falling apple to reach its terminal speed, *the apple tree would have to be taller than a skyscraper.* Here's why.

Note: These calculations are approximations! They ignore the effect of air resistance on the apple's acceleration.

Remember, speed = acceleration X time elapsed. Using this equation, the speed of an apple falling for 5 s is the product of the gravitational acceleration constant (10 m/s^2) times the time interval (5 s):

Speed = 10 m/s^2 X 5 s
= 50 m/s (speed 5 s after the stem breaks)

The terminal speed varies somewhat from apple to apple, depending on shape, skin condition, and other factors which influence its air resistance. It is safe to say that many apples would have a terminal speed which is considerably higher than 50 m/s. (Just for the sake of comparison, the speed of 50 m/s translates to about 112 miles per hour.)

But how far has the apple fallen in these 5 seconds, even though it has not reached its terminal speed during this time interval? The average speed of the apple is equal to its starting speed added to the last measured speed divided by 2.

$$\text{Average speed} = \frac{0 \text{ m/s} + 50 \text{ m/s}}{2} = 25 \text{ m/s}$$

Substituting this value in the formula which gives the distance fallen:

Distance = average speed X time interval
= 25 m/s X 5 s = 125 m

So, after falling more than the length of a football field, the apple still has not reached its terminal speed. Apple trees just do not grow tall enough!

◆Terminal speed: Does it change?

Objects falling through Earth's atmosphere are simultaneously acted upon by gravity and air resistance. As an object accelerates toward Earth, its air resistance (drag) increases. The force of gravity remains constant. At some speed, the *upward* force of air resistance will become equal to the *downward* force of gravity. When these opposing forces are equal, the object cannot continue accelerating. It begins falling at a constant speed called its terminal speed. It continues to fall at that constant speed until it hits the ground, or until the force of air resistance affecting it changes.

The terminal speed of an object is determined by a combination of factors including its shape, surface smoothness, and composition. Smooth shapes like bullets move through the air more easily than bulky, rough-surfaced objects like pinecones. The terminal speed of a falling streamlined object will be faster than the terminal speed of an irregularly-shaped object.

Since the shape of an object affects its air resistance, the terminal speed of an object can be changed by modifying its shape. For example, a free-falling skydiver wearing an unopened parachute reaches a terminal speed of over 50 m/s. If the parachute does not open, the skydiver will continue to fall at this constant rate of speed until he crashes into the ground.

But when the parachute opens, its surface area exposed to the air increases greatly. Because of its increased surface area, the upward force of air resistance acting upon the open parachute (and the skydiver) is much greater than before. By opening the parachute, the skydiver has effectively changed shape in such a way that his terminal speed decreases to less than 5 m/s. The skydiver's rate of fall decreases, and he can land without injury.

Opening the parachute changes the skydiver's terminal speed, but *does not* change his weight. The open parachute and skydiver weigh exactly the same as they did when the parachute was closed. Opening the parachute changes the *upward* force of air resistance, not the *downward* force of gravity acting on the skydiver.

READING 14

Gravity, Weight, and Weightlessness

◆Introduction

On Earth, weight is one of the forces that we often have to account for when examining action and reaction type motions, or when analyzing equal and opposite force pairs. But what *is* weight? How is weight related to gravity? How can astronauts be "weightless" even though their *mass*, defined in Module 1 as the "quantity of matter in the object," does not change? In order to begin to answer these questions, we need to know more about one of the fundamental forces of the universe: **gravitation**.

◆Gravitation

According to his own account, sometime around 1666 Sir Isaac Newton was sitting in an orchard. Seeing an apple fall to the ground, he suddenly realized that the acceleration of the apple toward the Earth occurred as a result of a force of attraction from the Earth to the apple, and a force of attraction from the apple to the Earth. The force, gravitation, exists between all objects in the universe. The force of gravitation is directly proportional to mass and inversely proportional to the square of the distance between the centers of the objects. The **gravitational constant, G**, which relates mass and distance to the amount of force being exerted, is thought to be the same everywhere in the universe.

Newton determined that the strength of the gravitational force exerted by a body (such as the Earth) upon another object (such as an apple) is determined by two factors:

1. the total amount of mass
2. the distance between the centers of the objects.

Newton's law of gravitation can be expressed mathematically:

$$F = \frac{GmM}{r^2}$$

Where,

G = universal gravitational constant

m = mass of one object

M = mass of second object

r = distance between the centers of the two objects

Based on this equation, two statements may be put forward:

1. As the amount of mass interacting increases, the total force exerted will increase.
2. As the distance between two masses increases, the force of attraction between them will decrease.

Experimental results support both of these statements. Small planets (planets that do not possess much mass) exert less gravitational force on objects at their surface than do large planets. For example, the Moon has less mass than the Earth, and exerts only one sixth as much gravitational force as the Earth. You would, therefore, weigh only one sixth as much on the surface of the Moon as you do on Earth.

The relationship between distance and gravitational force can be demonstrated by careful measurements performed on Earth. On Earth, for every 1000 m of altitude above sea level, the gravitational force that the Earth exerts decreases by 0.031 percent.

The decrease in gravity (and reduced air resistance at high altitude) probably assisted the record-breaking performances in the 1968 Olympic Games in Mexico City, which is 2300 m above sea level. The world record for the long jump at that time, 8.90 m, was set by Bob Beamon of the United States during that Olympiad. The record remained unbroken for 23 years. In 1991, Mike Powell, an American, finally set a new long jump record of 8.95 m.

The gravity of Earth is just one specific example of the universal phenomenon of gravitation. For predicting motion near the Earth's surface, Newton's law of gravitation can be stated in a form comparable to the equation for the second law of motion, $F = ma$. Gravitational force on Earth is expressed by:

$$F = mg$$

where g = the gravitational acceleration at the surface of the Earth.

The gravitational acceleration at the surface of the Earth is about 10 m/s^2. The metric unit of force, the newton, is the amount of force required to accelerate an object of 100 g (about the mass of a typical apple) at a rate of about 10 m/s^2.

◆What is weight? Are weight and mass different?

Weight is a way of representing the effect of the force of gravity acting on an object. Weight can be measured using a spring scale calibrated in newtons. Newtons are units of force. (An object weighing 1 N on Earth has a mass of about 100g.)

Another way to think about the concept of weight is that weight is proportional to the rate at which an object is being accelerated by gravity. On a planet less massive than the Earth, the object will accelerate more slowly, so its weight is less; on a more massive planet, it would accelerate at a greater rate, so its weight is more.

The mass of an object is determined by the total amount of matter contained in the object. (You may wish to review the three ways of defining mass presented in Module 1.) Moving from the Earth to the Moon *does not change the mass* of an object. A 1-kg mass would contain the same amount of matter on the Moon as it does on Earth; it *does not* lose 5/6 of its atoms when it reaches the Moon. That kilogram mass does, however, lose 5/6 of its *weight* on the Moon, because the gravitational attraction between the kilogram mass and the Moon is only 1/6 as strong as the attraction between the kilogram mass and the Earth.

◆Weightlessness: Is it possible to escape Earth's gravity?

Gravitational force decreases as a function of the square of the distance between two bodies. As the distance between the masses increases, gravitational force between them declines very rapidly. However, gravitation never truly reaches zero; everywhere in the universe there are small, perhaps imperceptible pulls of gravity between all the stars and planets and every other bit of matter in the universe.

Since gravitation exists everywhere in the universe, how can weightlessness occur? The answer lies in the type of motion that the "weightless" object (such as a spacecraft and its crew) is experiencing: *"weightless" objects orbiting Earth are undergoing free fall.*

◆How do satellites stay in orbit?

In his great work, *Principia*, first published over 300 years ago, Sir Isaac Newton explained how a satellite could orbit the Earth.

Using a diagram similar to the one on this page, he showed the motion of a cannonball fired from a high mountain top. The force of gravity acting on the cannonball would cause it to curve toward the Earth (path A). The greater the velocity of the projectile, the farther from the base of the mountain the projectile would land (path B).

If it were fired with sufficient velocity, the path of the cannonball curving toward Earth would exactly match the curvature of the Earth's surface (path C). The cannonball would continue falling toward Earth, but the path of its free fall would always remain parallel to the surface of the Earth. Assuming that there were no frictional forces (and that the cannon could be moved out of its way) the projectile would continue "falling around the Earth" in a circular path forever.

If a person were able to ride inside Newton's cannonball, what sensations would that passenger experience while orbiting the Earth in free fall? Think about the feelings you experience on a Ferris wheel or roller coaster. On both of these rides, when the car or chair begins falling after passing over the highest point of the ride, you begin to feel almost as though you are going to float out of your seat.

This sensation gives some inkling of the experience of true weightlessness. The roller coaster is not falling freely; it is accelerating rapidly toward Earth, and this motion cancels out *some* of the effects of gravity (as your stomach may tell you), allowing you to feel much lighter. Our mythical cannonball rider orbiting the Earth, however, really is experiencing free fall. She, the cannonball, and any objects she has with her become "weightless" while orbiting the Earth.

The three laws of motion are valid in the weightless conditions that she is now experiencing. On Earth, gravity is often one part of a pair of equal and opposite forces that affect moving objects. However, the effects of gravity have been canceled out by the free fall of the cannonball.

Everything inside is weightless, so many objects will move in ways that are quite different from what we have come to expect in the 1 *g* environment of Earth. However, when the absence of gravitational effects is accounted for, the behavior of objects in an orbiting spacecraft is exactly what Newton would have predicted.

◆Effects of weightlessness on moving objects

Living on Earth, we are so accustomed to the effects of gravity that canceling out the effects of gravity makes the motions of everyday objects seem strange. Our experiences lead us to expect the non-uniform, curving motion caused by gravity. Without careful analysis and accounting for all forces acting on moving objects, one might incorrectly conclude that the motions of everyday objects contradict the laws of motion.

In contrast, weightless objects orbiting Earth obviously do obey the first law of motion. Objects at rest (even objects resting in mid air) really *do* stay at rest when not under the influence of the unequal force, gravity. Moving objects really do move in straight lines in space.

The second law of motion predicts that the application of a force to an object will result in an acceleration. On Earth, in a 1 *g* gravitational environment, many forces we encounter are too small to overcome the forces of gravity and friction and produce accelerations. In the weightlessness of space, small forces do produce obvious accelerations.

Only one small group of people have ever had the opportunity to experience weightlessness for a prolonged period: astronauts participating in United States or Soviet space missions.

Astronauts train extensively under conditions of weightlessness. However, like everyone else who lives on Earth, astronauts tend to take the effects of gravity for granted when thinking about how different objects move.

Predicting how toys would perform in the weightless environment of the orbiting Space Shuttle required the astronauts to mentally cancel out the effects of gravity on these toys. They did not, in every case, recognize all the effects of gravity that occur on Earth. Therefore, the actions of some of the toys in space seemed strange to the astronauts, even though they knew that the weightless toys were obeying the laws of motion.

MATERIALS AND SOURCES

Guide for Teachers and Workshop Leaders

This section contains module-by-module master lists of materials, equipment, and audiovisual materials (if any) recommended for use in each activity, and gives ordering information about the audiovisual materials recommended for use with this manual. Substitutions for some materials may be suggested in the "Preparation" section of each activity.

Master Lists of Materials

◆Module 1: Sizing it Up

Master list of materials and equipment for each group:

1 meter stick
1 paper clip
2 2-ounce fishing sinkers (or similar-size masses)
1 1-ounce fishing sinker (or similar-size mass)
1 clock or watch with a second hand
1 empty bottle of dishwashing liquid and its push-pull top
1 container of water (to fill the detergent bottle)
1 metal pie pan
optional: a ring stand and clamp
The following may be used by the entire class or workshop group:
1 ball of string
heavy-duty scissors

◆Module 2: Mass and Force

Master list of materials and equipment for each group:

a 2 x 4 board about 46 cm in length
3 nails
1 hacksaw blade
a ruler or wood molding 46 cm long
a bolt, washer, and nut
1 styrofoam cup
6 2-ounce fishing sinkers (or similar-size weights)
1 clock or watch with a second hand
1 metric ruler or meter stick
safety goggles for each person
1 heavy-duty rubber band
1 ring stand with clamp
2 paper clips
1 sheet of graph paper for each person
optional: a spring scale calibrated in newtons (or grams)
The following may be used by the entire class or workshop group:
duct tape
optional: 1 ball of string

Audiovisual equipment:

videocassette player (VHS format) and television
Eureka! video series programs #1 and #2

◆Module 3: Constant Speed Versus Acceleration

Master list of materials and equipment for each group:

1 meter stick
1 battery-operated electric model car
1 clock or watch with a second hand
1 push-pull top from a bottle of liquid detergent
200 ml of water with food coloring added (100 ml used in each of two activities)
1 medicine dropper
1 model car that rolls freely
a table with smooth surface about 2 meters long (or a sheet of plywood or wide board of similar length)
several bricks or old books for propping up one end of the table
1 index card (3" x 5" or larger)
1 1-ounce lead fishing sinker (or similar-size weight)
1 ruler or stick at least 30 cm long
a felt tip pen
10 "split shot" sinkers
an aluminum pie pan or cookie sheet
pliers

The following may be used by the entire class or workshop group:

1 ball of string
masking tape
rags, paper towels, or sponges for cleaning up
sewing thread

Audiovisual equipment:

videocassette player (VHS format) and television
Eureka! video series program #5

◆Module 4: Interactions of Force, Mass, and Acceleration

Master list of materials and equipment for each group:

1 meter stick
2 sections of N-gauge model railroad track
a grooved ruler
5 marbles of the same size
several books for supporting the ruler and tracks
1 slider car (model provided in activity)
1 Stomper™ car (or similar motorized model car)
1 model railroad car
60 2-ounce lead fishing sinkers (or other objects of similar mass to serve as cargo)
1 475-mL plastic cup
1 timer or clock with a second hand

The following may be used by the entire class or workshop group:

1 ball of string
masking tape

Audiovisual equipment:

videocassette player (VHS format) and television
Eureka! video series programs #3 and #4

◆Module 5: Applying the Laws of Motion

Master list of materials and equipment for each group:

safety goggles for each person
1 meter stick
3 empty Kodak™ 35-mm film containers
4 Alka-Seltzer™ tablets
2 2-ounce fishing sinkers
1 section of N-gauge model railroad track 70 cm long
1 100-ml container of water
20 pieces uncooked fettuccini
a heavy book
1 ping-pong ball
a package of plasticine-clay type paper adhesive such as Holdit
1 electric hair drier
optional: clamp to hold hair drier in position
The following may be used by the entire class or workshop group:
rags, paper towels, or sponges for cleaning up
masking tape

◆Module 6: "Hidden Forces" Affecting Motion

Master list of materials and equipment for each group:

a 2 x 4 board about 35 cm in length
1 cup hook
a spring scale calibrated in newtons (or grams)
a 500-g mass (a 1-pound can of soup, lead sinkers totaling 16 ounces, or other similar objects may be used)
4 or more *round* pencils or sections of dowel
1 meter stick
1 timer which can measure 0.1-s intervals
1 2-cm length of yarn
1 penny
1 hardcover book
The following may be used by the entire class or workshop group:
masking tape
1 paper airplane
1 Slinky™
1 yo-yo
1 set of jacks and a ball
1 Wheel-lo™

Audiovisual equipment:

videocassette player (VHS format) and television
Eureka! video series programs #6 and #7
Toys in Space videotape

Recommended Audiovisual Materials

Depending on the needs, interests, and abilities of your students, these audiovisual materials may provide useful supplements or alternatives to textbook or lecture presentations of selected concepts of Newtonian mechanics.

◆*Toys in Space* videotape

This videotape of astronauts exploring the behavior of simple toys in a weightless environment was made during a Space Shuttle mission. Eliminating the effects of gravity (a pervasive unequal force on Earth) produces some unexpected results when the toys are tried out in orbit. Activity 27 includes commentary on the videotape.

For more information contact:

Informal Science Study
University of Houston
Room 112 Farish Hall
Houston, Texas 77004
(713) 749-1692

The videotape may be obtained for $29.95 plus $3 shipping and handling from:

Sopris West Incorporated
1140 Boston Avenue
Longmont, Colorado 80501
(303) 651-2829

◆*Eureka!* video series

Eureka! is an animated video series produced by TVOntario (© 1981). It uses examples drawn from everyday experiences to demonstrate the behavior of matter in motion. The cartoons are useful for introducing or reviewing the concepts of mechanics.

Many school districts already have copies of *Eureka!* available for use. Check with your district or state instructional TV personnel first to see if yours is one of them.

For information on how to rent or purchase copies of *Eureka!* contact:

TVOntario
Suite 308
1140 Kildaire Farm Road
Cary, North Carolina 27511
(800) 331-9566 or (919) 380-0747 (in NC)

Metric Conversions

Only two countries in the world, Burma and the United States, persist in using the English system of measurement.* Although some U.S. industries are beginning to build items to metric standard measurements in order to become more competitive in international markets, the old units of feet, pounds, quarts, and other English measures are still commonly used for many consumer items in the United States. Until we join England in giving up the English system, converting from English to metric units or vice versa will sometimes be necessary. The following table tells how to do the most common conversions:

Rules for English to Metric Conversions

Multiply	by	this number		to get
_____inches	X	2.54	=	_____centimeters
_____feet	X	0.3048	=	_____meters
_____miles	X	1.609	=	_____kilometers
_____quarts	X	0.945	=	_____liters
_____gallon	X	3.7	=	_____liters
_____pounds	X	0.453	=	_____kilograms
_____°F – 32°	X	0.556	=	_____°C

Popular Science, Vol. 233, No. 1. (July 1988, p. 45, "What's News")

METHODS OF MOTION

Glossary

Acceleration: The rate at which an object speeds up or slows down.

Accelerometer: A device used to measure acceleration or determine if acceleration is present (acceleration meter).

Air resistance: A force exerted on a moving object opposite to its direction of motion due to the friction between the object and air. Air resistance is also called *drag* or *air friction*.

Calibrate: To adjust a measuring instrument so as to reduce error in measurement; to standardize the instrument by determining its deviation from an accepted value.

Constant speed: An object is traveling at constant speed if it moves the same distance within every equal time interval.

Control: An experiment (or factor in an experiment) that remains unchanged for purposes of comparison. The control allows an experimenter to judge the effects of changing factors in an experiment.

Cycle (of a pendulum): One of the "out and back" swings of a pendulum.

Delta notation: A symbolic indication of an interval using the Greek letter delta, written Δ ; for any quantity x, Δ x means "the change in x" or "calculate an interval of x."

Displacement: The distance that an object moves.

Dynamics: A branch of mechanics that concentrates on the study of bodies in motion.

Error in measurement: The difference between an observed or calculated value in an experiment and the true value. An error in this sense is not a mistake, but a variation from the true value present to some degree in all experiments.

Force: A push or pull in a particular direction that can be applied to an object.

Friction: Resistance to motion between two bodies in contact.

Gravity: The force of attraction between all objects in the universe. Wherever there is mass, there is gravity.

Hypothesis: An educated guess that is based on available information and experience. It is tested by scientific experiment.

Inertia: A property possessed by all matter that can be thought of as laziness or "difficult-to-moveness"; it is the tendency of matter to keep doing what it is already doing. Inertia is the subject of Newton's first law, which states: an object at rest tends to stay at rest, and an object in motion tends to stay in motion in a straight line and at a constant speed unless acted upon by an unequal force.

Inertial balance: A device used to measure mass by making use of the relationship between mass and inertia.

Mass: A property of matter related to inertia. As the mass of an object increases, so does its inertia. Mass can be thought of as the quantity of matter in an object.

Mechanics: The branch of physical science that describes the behavior of bodies in motion; mechanics deals with energy and forces and their effects on bodies.

Momentum: The product of an object's mass and velocity; Newton called momentum the "quantity of motion." It can be thought of an an object's "bashing power."

Newton: A unit used in measuring force. 1 newton is the amount of force required to accelerate a 1-kg mass 1 m/s^2.

Operational definition: A non-mathematical definition of a term that includes an operation to perform and observations to make in order to determine whether the defined phenomenon is present.

Pendulum: Any object attached to a fixed point so that it swings freely back and forth under the action of gravity.

Period (of a pendulum): The time it takes a pendulum to swing out and back to its point of release.

Speed: The rate of change of position; speed combines information about how far an object travels with how long it takes to travel that distance.

Vacuum: A space devoid of all matter, including air.

Variable: A factor that is changed in an experiment in order to see how it affects the results of the experiment.

Weight: On Earth, a force due to the gravitational attraction between an object and the Earth. Weight is a downward force that acts on the object.